Columbia University

Contributions to Education

Teachers College Series

No. 405

AMS PRESS
NEW YORK

A STUDY OF PROBLEM MATERIAL IN HIGH SCHOOL ALGEBRA

BY

JESSE JEROME POWELL, Ph.D.

INSTRUCTOR IN MATHEMATICS
THE COLLEGE OF THE CITY OF NEW YORK

Teachers College, Columbia University
Contributions to Education, No. 405

Bureau of Publications
Teachers College, Columbia University
NEW YORK CITY

1929

Library of Congress Cataloging in Publication Data

Powell, Jesse Jerome, 1891-
A study of problem material in high school algebra.

Reprint of the 1929 ed., issued in series: Teachers College, Columbia University. Contributions to education, no. 405.

Originally presented as the author's thesis, Columbia.

1. Algebra--Study and teaching. 2. Algebra--Problems, exercises, etc. I. Title. II. Series: Columbia University. Teachers College. Contributions to education, no. 405.

QA159.P6 1972 512.9'0076 71-177162
ISBN 0-404-55405-9

Reprinted by Special Arrangement with Teachers College Press, New York, New York

From the edition of 1929, New York
First AMS edition published in 1972
Manufactured in the United States

AMS PRESS, INC.
NEW YORK, N. Y. 10003

ACKNOWLEDGMENTS

Among those who assisted in this study especial acknowledgment should be made to Professor C. B. Upton for his valuable and constructive criticism; to Professor H. A. Ruger for suggestions concerning the statistical analysis of data used; and to Professor W. D. Reeve for his cognizance of present-day trends and needs in education which led him to collect over a number of years the material which has made this investigation possible. Professor Reeve's scholarly, human interest has been invaluable throughout the undertaking.

The author gratefully remembers his wife, Eula Mitchell Powell, for her coöperation in tabulating data and reading the manuscript.

J. J. P.

CONTENTS

DIAGRAMS

A STUDY OF PROBLEM MATERIAL IN HIGH SCHOOL ALGEBRA

CHAPTER I

REAL AND GENUINE PROBLEMS IN ALGEBRA

Introduction

Efforts now being put forth to educate all of our youth for the best participation possible in citizenship have made inevitable a revision in the objectives in the teaching of mathematics. In turn, these new objectives require a reassortment of the material through which they may be attained.

Mathematical theories and processes have changed only through a process of natural development, but there has risen on the part of the general educator, as well as the mathematician, a demand for genuine problems which will show the application of these theories and practices.

Not in opposition to, but perhaps in addition to, this economic demand for genuineness in problem material comes the demand on the part of the pupil for problems that are real to him and that will arouse his interest.

The chief aim of this study has been to compile a list of five hundred of the best problems in elementary algebra that are available at the present time. This selection has been made in the light of recent advances in the aims, objectives, and teaching practices in the education of the youth of the country. It is based upon expert opinion, secured from a group of experienced, well-trained, and widely distributed teachers of high school mathematics, with reference to a large number of possible choices, and also upon the preferences which a large number of high school pupils have shown for these problems.

Definition of Terms

The following terms which characterize these problems are defined as follows:

1. *Genuineness.* As herein used, a problem is genuine if anyone in any place might need to solve it at present.

2. *Importance.* The importance of a problem depends upon the frequency with which it might occur in a life situation.

Genuineness and importance are based upon their application to situations in the lives of the greatest number of individuals in general. The primary aim in high school education is training for citizenship. The material that makes up the courses for this training should be selected, therefore, on the basis of its usefulness to the greatest number of individuals rather than on its use in the training of experts. Training of experts should be left largely to the technical or trade school.

3. *Reality or Interest.* A real problem is one that issues a definite challenge to the pupil. It is one that he enjoys attacking and working through.

4. *Difficulty.* Difficulty, as used in this study, is based upon the percentage of pupils solving the problem. It is agreed that a problem which the pupils fail to solve might be one of the easiest to solve if the principles were properly taught.

These definitions are later stated more fully in the instructions to teachers and to students under the section, Methods of Procedure.

The Problem Restated

In the light of the above definitions, the problem might well be restated in the form of questions as follows:

1. What would be the best types of problems to put into a list of four or five hundred problems which is to be the material used in teaching high school algebra?

2. What types of problems would be most genuine, and, of these, which would be considered most important by experts in the field?

3. Which of these genuine and important problems possesses a fairly high degree of reality to the student?

4. Other phases of the problem which grow out of these main points are:

(1) According to the evidence obtained, what relationships exist between genuineness and importance; between importance and interest as estimated by the teacher; between importance and interest as evaluated by the pupil; between the interest of the pupil and the estimated interest of the teacher; between pu-

pil's interest and difficulty; between the teaching of a problem and the teacher's estimates of interest?

(2) What are the elements that make problems interesting or uninteresting to pupils in general?

Source of Material

The problems that serve as the basis for this study, and from which the lists that go to make up the main body of the thesis were taken, were selected from a list of 3,000. These problems were contributed by graduate students from every section of the United States and from many other countries, assembled in classes in Teachers College, Columbia University. These students, under the direction of Professor W. D. Reeve, were asked to submit problems that are original as far as possible, illustrating varying degrees of genuineness and importance. When a problem was copied from a textbook, the author and the text were to be indicated. The copied problems were used in the study to determine relationships, but an honest effort has been made to eliminate them from the final lists entered in this study. Many of the problems, though original so far as the data are concerned, were necessarily patterned after types found in textbooks.

Procedure in General

The primary purpose of this study is the compilation of the best possible lists of elementary problems in algebra. After the tabulation and analysis of data were made, however, the relationships existing between the factors, *interest*, *genuineness*, *importance*, and *difficulty* in algebra problems seemed worthy of a statistical report. This statistical analysis has been made in the hope of pointing the way to further and more extended study, so that, as a result, still better material may find its way into our textbooks. This report is found in the Appendix.

CHAPTER II

METHOD OF PROCEDURE

Problem Types

Algebra problems may be divided into a number of "types." As a working basis, the following types have been used: age, average, cost, digit, formula, graph, indirect measurement, investment, insurance, lever, mensuration, mixture, number, percentage, permutation and combination, proportion, rate, puzzle, earning and saving, sharing and work.

Collecting and Tabulating Data

The only basis for the selection of the problems out of the list of 3,000 collected was the elimination of exact duplicates and the inclusion of a number of all the different types. Upon this basis, a list of 1,300 problems was selected and divided into groups of approximately 150; each group contained a number of every type.

Enough duplicate problems were distributed through each group to serve as a check on the individual as to sincerity in rating. If there was wide variation in the evaluation of these duplicate problems, it could be concluded that the individual was marking only at random. Thus it was possible to safeguard against the tabulation of worthless data.

These groups of 150 problems were mimeographed and given to 500 experienced and trained teachers of algebra. These teachers were also members of classes at Teachers College, Columbia University, and came from all sections of the United States and from other countries. The following mimeographed sheet of instructions was given to each person with each list of problems:

I. Academic Preparation and Experience.

1. Degree or Diploma held......Institution Issuing......Date......
....
....

2. Place a check before each of the following courses that you have had
......College Algebra......Analytics......Projective Geometry
......Descriptive Geometry......Differential and Integral Calculus
......Differential Equations

3. Have you had courses previously in the Teaching of Algebra?......

4. How many years have you taught High School Algebra?......

II. INSTRUCTIONS.

In the spaces indicated, rank the following list of problems on the basis of:

1. *Genuineness.* Genuineness is here used to indicate the probability that a problem has of occurring in an actual life situation.

If you think that a problem would never occur in a life situation mark it zero (0); if you think that it might occur in every detail as it is written, mark it ten (10); mark all the others on the scale of 0 to 10 according to your best judgment of its genuineness.

2. *Importance.* Importance is used here to designate the frequency of occurrence of those problems that do arise in life situations. If you think that a problem would occur not more than once in five years, mark it 0; if you think that it would occur at least once every month, mark it 10. Rank the others on the scale of 0 to 10 according to your best judgment of its importance.

3. *Interest.* If, from your experience and best judgment, you think that the solving of a problem would be *disliked* by all of any class of pupils, mark it —10; if you think the solving of a problem would be enjoyed by all of a class, mark it 10. Zero interest may be defined as neutral interest. If you think that the solving of a problem would create neither interest nor lack of interest or that it would make little or no difference to a class mark it 0.

4. In marking these problems, think of their application to the greatest number of citizens and not to specialists in some field.

In tabulating the results, the opinions of an individual were rejected if he lacked at least three years of experience as a teacher of high school algebra or training including differential and integral calculus. Also, the scoring by one whose work lacked the evidence of honest effort was rejected. A final count gave an average of 250 opinions that could be considered expert on each type of problem.

Of the 1,300 problems submitted to teachers for evaluation, 200 thought suitable for the ninth grade were selected from those rated 5 or above, on a scale of 0 to 10, in genuineness and importance. In addition to those rated high in genuineness and importance, there were selected some of all types, whether they ranked high in these two factors or not. This was done in order to determine if possible the elements that go to make up problems of varying degrees of interest.

These problems were mimeographed and sent out with the following instructions to pupils who were just completing ninth year algebra:

Name of Pupil........................School........................

City ....State ....

Grade ....Instructor ....

INSTRUCTIONS.

1. Read each problem carefully, and then set up equations for its solution in the space beneath it, as shown in the example at the bottom of this page. Do not solve the equations.

2. After you have formed the equations for a problem, make a check in one of the spaces, indicating whether you like the problem "very much," "fairly well," or "not at all" as shown in the example below. If you cannot form the equations, check each problem for "interest" anyway.

EXAMPLE.

John owes in all $24 to his father and older brother. If he owes the father twice as much as the brother, how much does he owe each?
Like very much...... Like fairly well...... Do not like......
Solution: Let $x =$ the amount John owes the brother
and $2x =$ the amount he owes the father
Then $x + 2x = \$24$

The results from this work were tabulated under the headings, *interest* and *difficulty.*

From the data gathered through these investigations, the means and the standard deviations were calculated as measures of central tendencies and of dispersions, respectively. The correlation coefficients with their probable errors were computed to serve as a basis of analysis. As a check against these measures, association is shown by means of percentages and contingency coefficients. Associations are shown between the following factors:

1. Genuineness and estimated interest by the teacher.
2. Importance and estimated interest.
3. Estimated interest and interest as evaluated by the pupils.
4. Pupils' interest and importance.
5. Pupils' interest and difficulty.

CHAPTER III

GENUINENESS, IMPORTANCE, INTEREST, AND DIFFICULTY IN ALGEBRA PROBLEMS: A STATISTICAL ANALYSIS

In considering the relationships between genuineness, interest, importance, and difficulty in algebra problems, a number of questions present themselves: (1) To what extent are the problems that are genuine considered of any great importance to the average citizen? This question would probably be more lucid if it were stated: To what extent, in the light of evidence given by the group of experienced teachers used in this study, should the average citizen be able to solve these genuine problems? (2) How much interest do teachers of algebra think pupils should attach to problems that are considered genuine or important? (3) Do importance and difficulty have any influence upon the interest of the pupil? (4) Is the same degree of interest that is estimated by the teacher expressed by the pupil?

The reliability of any kind of measurement of relationship between the above attributes depends largely upon the degree of accuracy with which each can be located according to a scale of zero to ten. In this study this method of scaling has been employed in order to secure as far as possible objective measures of the more or less subjective attributes.

The results that are shown by these data are not intended as absolute measures, but it is believed that they do represent significant tendencies.

Genuineness and Estimated Interest

In order to answer the question, "How much interest do teachers think pupils should show in problems that are considered genuine or important?" there must be established some kind of measurement of relationship. As was previously pointed out, this necessitates the measuring of relationships between qualitative attributes.

For the purpose of tracing the association between genuineness and interest as estimated by the teacher, the data were classified in Table I in a 2 x 2-fold form.[1] The problems rated from 0 to 4, inclusive, are classified *not genuine* or *low estimated interest,* and those rated from 5 to 10, inclusive, are classified as *genuine* or *high estimated interest.*

TABLE I

GENUINENESS AND ESTIMATED INTEREST

	Genuine	Not Genuine	Total
High estimated interest ...	863	29	892
Low estimated interest ...	229	87	316
Total	1,092	116	1,208

Calculating the proportions of high and of low estimated interest in the genuine problems from this table, we have

Proportion of genuine problems with high estimated interest	863/892 = 97	per cent
Proportion of genuine problems with low estimated interest	229/892 = 25.6	per cent

These ratios indicate that the association is positive. For comparison we are enabled to deduce a tetrachoric correlation[2] from this same table, $r_t = 0.61 \pm .027$. Thus it may be concluded that there is distinct positive association between these attributes.

Importance and Estimated Interest

Here, as in the case of genuineness and estimated interest, the same problem presents itself, that of measuring the relationship between qualitative attributes. In order to trace the association between these two variates, we may again form a tetrachoric

[1] Yule, G. U., *An Introduction to the Theory of Statistics,* Chap. III. Chas. Griffin & Co., London, 1924.

[2] Pearson, K., "On a Novel Method of Regarding the Association of Two Variates," *Draper's Company Research Memoirs,* Biometric Series, VIII. Dulau & Co.

table similar to Table I by classifying the problems rated from 5 to 10 as *important,* or as *high estimated interest,* and those rated from 0 to 4, inclusive, as *not important,* or as *low estimated interest.*

TABLE II

IMPORTANCE AND ESTIMATED INTEREST

	IMPORTANT	NOT IMPORTANT	TOTAL
High estimated interest ...	682	186	868
Low estimated interest ...	98	234	332
Total	780	420	1,200

From Table II, it may be observed that

Proportion of important problems with high estimated interest .... 682/868 = 80 per cent

Proportion of important problems with low estimated interest .... 186/868 = 21 per cent

These ratios indicate that the association is positive and well marked. The tetrachoric correlation ($r_t = 0.637 \pm .02$) bears out this observation, and, again, it may be concluded that there is distinct positive association between genuineness and estimated interest.

Other measures of relationship may be applied to the same data and the measures compared and checked against one another.

If the data are classified in a 4 x 4-fold form, the contingency coefficient[3] is found to be 0.52, showing a rather close association of estimated interest and importance in algebra problems.

The above measurements of association and the coefficient of correlation with its probable error ($r \pm \text{P.E.} = 0.66 \pm .01$)[4] serve as a double check on the interdependence of these two traits. A more complete picture of this relationship may be had from the regression equation.[5] It is found that the regression of interest (y) on importance (x) is represented by the equation

[3] Yule, G. U., *An Introduction to the Theory of Statistics,* pp. 64 ff.

[4] Computed from Table III.

[5] *Op. cit.,* Chap. IX.

$y = .73x$ and that of importance x on interest y by the equation $x = .59y$. In Diagram 1 the regression lines, as always, are drawn through the intersection of the lines of the means of importance ($M_x = 5.1$) and of interest ($M_y = 5.6$). The standard deviation (σ_x) around the mean of importance is 2.1 and that around the mean of interest (σ_y) is 2.4. The regression of interest on importance ($y = .73x$) is represented by the line *AB*, and the regression of importance on interest ($x = .59y$) by the line *CD*.

The standard error of estimate for the values of estimated interest and importance ($\sigma_{y \cdot x}$) is 1.8. The probable error of estimate, P.E. (est. y), indicates that the chances are about 50 in 100 that the actual estimated interest does not differ from the predicted by more than 1.2.

TABLE III

Correlation Between Importance and Interest Estimated by the Teacher, Related to Problems in High School Algebra

Estimated Interest	Importance 0	1	2	3	4	5	6	7	8	9	10	Total
10	....	....	....	2	3	15	8	9	18	7	12	74
9	....	....	....	....	5	11	5	13	15	13	6	68
8	1	1	1	3	6	35	31	36	38	9	2	163
7	....	1	3	5	5	37	24	29	25	6	3	138
6	1	1	6	14	13	47	21	20	13	....	....	136
5	4	10	16	45	40	108	38	21	10	1	....	293
4	1	5	8	15	28	38	8	5	2	....	....	110
3	6	3	14	33	28	26	3	....	1	....	....	114
2	4	3	12	17	13	5	1	....	....	....	....	55
1	....	4	9	6	1	4	....	....	....	....	....	24
0	10	....	7	7	1	2	....	....	....	....	....	27
Total .	27	28	76	147	143	328	139	133	122	36	23	1,202

Importance and Public Interest

In order to obtain their evaluations as to interest in problems, the pupils were asked to score a problem according to whether

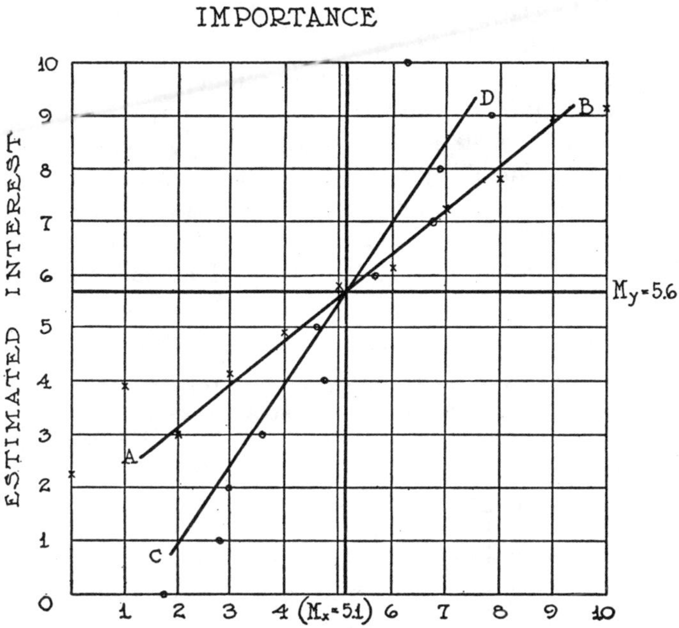

DIAGRAM I

CORRELATION OF IMPORTANCE AND PUPIL INTEREST

they (1) liked it very much, (2) liked it fairly well, or (3) did not like it at all. In calculating the degree of association between the traits, importance and pupil interest, a division of Table 3 into a 3 x 3-fold form was made. The evaluations of the teachers as to importance were divided into three classes; those ranked from 7 to 10, inclusive, were classified as very important; those ranked from 4 to 6, inclusive, fairly important, and those ranked from 0 to 3, not important at all. From this division of the data the contingency coefficient was found to be zero.

By computing the mean value of the pupils' ratings, their evaluations as to interest were distributed on a scale of 0 to 10, with class intervals of 1. This distribution was made for the purpose of calculating the coefficient of correlation.

Table IV exhibits the frequencies of the pairs of values, as to importance and pupil interest, lying within each class interval. The correlation coefficient obtained from this table ($r = -0.03$) was practically zero with a probable error of 0.04. Hence the contention can be made that there was a definite tendency toward low correlation between importance of problems and interest expressed by the pupil.

The regression of importance (y) on interest (x) is shown by the equation $y = -0.04x$, and that of interest x on importance y by the equation $x = -0.02y$. Equation (1) is represented in Diagram II by the line AB; and equation (2) by the line CD. The mean of importance (M_y) is 5.5, and the mean of interest (M_x) is 4.2. The standard deviation around the mean of importance (σ_y) is 2.0 and that around the mean of interest (σ_x) is 1.78.

TABLE IV

Correlation Between Pupil Interest and Importance, Related to Problems in High School Algebra

		Pupil Interest											Total
		0	1	2	3	4	5	6	7	8	9	10	
Importance	10	....	....	....	....	....	....	....	....	....	....	....	....
	9	....	....	....	....	....	2	4	....	....	....	....	6
	8	....	2	4	10	2	10	4	4	....	....	....	36
	7	....	4	....	6	6	4	2	....	2	....	....	24
	6	....	6	4	2	4	6	4	6	....	2	....	34
	5	....	4	10	6	4	12	6	4	....	....	....	46
	4	....	4	....	2	4	4	....	....	....	....	....	14
	3	....	2	....	4	4	6	2	....	....	....	....	18
	2	....	....	2	....	....	2	....	....	....	....	....	4
	1	....	....	....	....	....	....	....	....	....	....	....	....
	0	....	....	....	6	2	....	4	4	2	....	....	18
Total		....	22	20	36	26	46	26	18	4	2	....	200

It will be noted from Diagram II that the interdependence of importance and pupil interest is slight.

Pupil Interest and Interest as Estimated by the Teacher

In tracing the association of interest expressed by the pupil and that which he is supposed to show according to the estimates of the teacher, the same procedure was followed as in the last

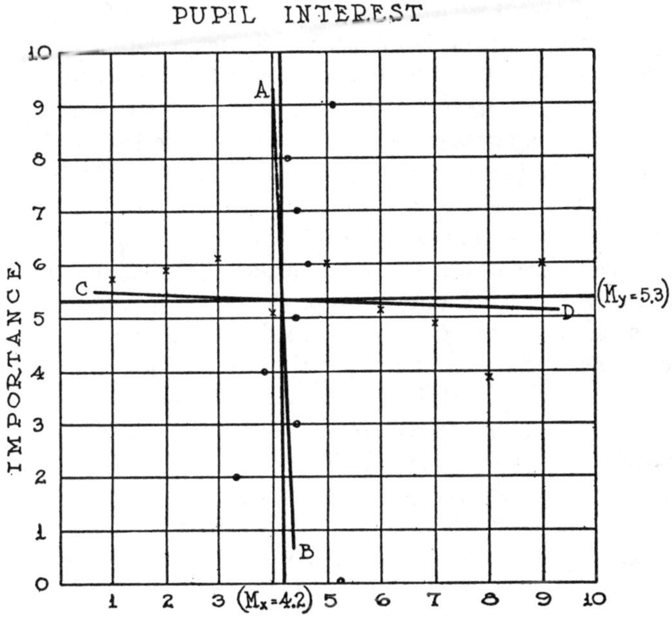

DIAGRAM II

CORRELATION BETWEEN IMPORTANCE AND PUPIL INTEREST

case. As would be expected, the contingency coefficient is low (0.17). The correlation coefficient is correspondingly low ($r =$ 0.11) with a rather large probable error (P.E. = 0.05). Again, the contention can be made that there is practically no association between interest as expressed by the pupil and as estimated by the teacher.

DIFFICULTY AND PUPIL INTEREST

The expression, "Pupils do not like to work problems that are difficult," is somewhat discounted by a comparison of the results obtained from different large groups. One group of pupils to which the list of 200 problems was given had not been taught a large percentage of the types represented. Rated on a scale of difficulty of 0 to 100, according to the number missed, the problems ranged from 57 to 98 in difficulty with a mean difficulty of 92 and a standard deviation of 8.8. The mean value of interest was found to be only 3.2, with a standard deviation of 1.6. The coefficient of correlation, though negative, was found to be significant (—0.51) with a probable error of 0.035. As is shown by the probable error, this indicates a rather high degree of negative association between interest and difficulty.

A different result is obtained from a second group of pupils. In this case the types of problems contained in the list had been taught. Here the range of difficulty was from 20 to 70, the mean being 45 with a standard deviation of 12.8. The mean interest value was found to be 5.1 with a standard deviation of 1.9. The correlation, although negative, is rather low ($r =$ — 0.18) with a probable error of 0.068.

As a check on the correlation coefficient, a contingency coefficient was computed. Dividing the table in a 3 x 3-fold form and calculating from this, it was found to be 0.27.

A comparison of the results shows that in the case of a high percentage of new or very difficult problems, the negative correlation of interest with difficulty is comparatively high; in the case of problems with which pupils are more familiar, the correlation between interest and difficulty is strikingly low.

Conclusions

This analysis warrants the following conclusions:

1. There seems to be a tendency on the part of teachers to think that if a problem sounds plausible the pupils should take a reasonable amount of interest in its solution. There is a still greater tendency to think that if the average citizen is likely to be called on frequently to solve a problem of a certain type, the pupil should show great interest in it.

2. The pupils' reactions are different from those of adults. In a great number of instances the pupil cares little whether a problem is genuine or important. He may be brought to accept and solve problems that are important in the life of the average citizen, but it does not necessarily follow that he will like them. The contrary is more likely to be the case.

3. The results obtained from this study indicate rather conclusively that the teacher's choice is not a sufficient basis for selection of problems with reference to interest. In the majority of cases the pupils report little concern in problems rated by the teacher as highly interesting.

4. There is some negative correlation between difficulty and interest on the part of pupils. The indications are, however,

that most pupils are not materially disturbed by the introduction of new or difficult problems. In the cases studied, some of the problems that were rated highest in interest by the pupils were found to be among the most difficult (difficulty being based on the number missing). Discretion, however, should be used in the number of new situations introduced at a time. Interest would likely be destroyed if all of a group of problems were difficult.[6]

[6] Thorndike, E. L., *The Psychology of Algebra*, pp. 409 ff. Macmillan Company, New York, 1924.

CHAPTER IV

TYPES OF GENUINE AND IMPORTANT PROBLEMS IN ALGEBRA: AN ANALYSIS

The Reorganization Movement

An inspection of mathematics textbooks written within the last ten years and of recent literature on the subject of teaching mathematics gives evidence that "what to teach" is a fundamental problem of to-day. Mathematics departments of our schools are attempting to solve this problem, they wish their work to become more effective, and, if necessary, they are willing to reorganize.

The movement to reorganize mathematics in secondary schools was instigated about 1900 by such men as Perry, Klein, and Moore. It assumed definite form in America with the organization of the National Committee on Mathematical Requirements. The report [1] of this committee has had tremendous weight throughout the country in the reconstruction of curricula in mathematics for secondary schools.

One phase of this movement for reorganization has been concerned with the elimination of what is now considered obsolete, irrelevant, and otherwise undesirable problem material. There is coming to be a concensus among educators in general that in attaining certain aims and objectives in education, it is much better to use material that has a direct application in the lives of individuals. This policy is sound, especially if this material will serve as well as any other in attaining such aims and objectives.

A list of illustrative examples will serve to show the types of problems that are considered by experienced teachers of high school algebra to be most worthwhile to a majority of individuals.

[1] *The Reorganization of Mathematics in Secondary Education,* United States Bureau of Education, Bulletin No. 32, 1921.

Types of Important Problems

The following list of problems contains samples of all types. For the purpose of comparison and analysis, a number of problems with varying degrees of importance is included in each type. The number in parentheses at the end of each problem indicates on a scale of 0 to 10 the degree of its importance.

1. *Graphs.*

(*a*) A man 20 years of age may have his life insured with a certain company by payment of $18.50 per year per $1,000 of insurance. The following table gives the amount that must be paid if the insurance is taken at other ages.

Age	Payment	Age	Payment
25	$20.10	45	$37.80
30	22.92	50	46.65
35	26.55	55	58.84
40	31.33	60	75.77

Construct a graph to show the relations between the payments and the ages of the persons when insured. Estimate from your graph the payment necessary to insure a person aged 27 for $1,000. (9)

(*b*) An automobile starts from Chicago and travels west at the rate of 30 miles per hour. Two hours later a second car starts from the same point and travels the same route at a uniform rate of 40 miles per hour. Construct a graph showing when the second automobile overtakes the first, and how far they have traveled. (6)

(*c*) Compare by means of a graph the enrollment of your school for the past 10 years with the valuation of property of the city during that same time. (5)

Graphs rank high in importance, not because of the importance of the situation they represent, perhaps, but because of their value in picturing the dependence of one variable on another. It is interesting to note from these examples that the one given the highest value describes an economic situation.

2. *Formulas.*

(*a*) Represent the following statement by a formula: The interest is equal to the principle times the rate times the time. Find the interest on $650 at 6% for five years by using your formula. (10)

(*b*) In a certain factory, employees receive 52¢ per hour for some work, $2\frac{1}{2}$¢ a piece for piecework, and an extra $\frac{1}{2}$¢ for each piece over 100. Set up a formula to show the daily wages of a person who works both by the hour and the piece. Find his wages if he works $3\frac{1}{2}$ hours, and during the remainder of the day finishes 124 pieces. (5)

(*c*) When a ship springs a leak, the quantity of water, in tons per hour, that flows in is given by the formula:

$$W = A(20d)^{1/2}$$

where

A = the area of the hole in square inches.

d = the depth of the hole in feet below the surface. (0)

A large percentage of the formulas receive high rating in importance. Those that are ranked low describe situations in which the average citizen has little interest or situations that only the expert would need to know about, as indicated by example (*c*). As in the case of the graph, the problem that is given the highest value is one that expresses economic relations.

3. *Averages.*

(*a*) If my semester average is obtained by counting the class work as two-thirds and the final examination one-third, what final examination must I make in order to pass the course with a grade of 70 when my class work is worth only 64? (9)

(*b*) A tennis tournament was held between two schools. The results were such that one school won two more games than the other. If the total number of games played was 8, what were the scores of each team? (5)

4. *Indirect Measurement.*

(*a*) How would you measure indirectly the width of a river? (7)

(*b*) Henry observed that after letting out 500 feet of his kite string his kite was directly over a near-by residence. He measured the angle of elevation of the kite and found it to be 30 degrees. How high was the kite, and how far away was the residence? (5)

Average and indirect measurement problems are generally given high values. As in the case of averages, these problems have direct application in the life of the pupil as shown in example 3(*a*), or they describe situations in which anyone may find himself, as in the measuring of some inaccessible object described in 4(*a*) or 4(*b*).

5. *Investment.*

(*a*) How much money would your father have to invest at 4% interest, compounded semiannually, to earn every year what he uses to keep his family for a year? (9)

(*b*) A man has $3,000 to invest. He has a chance to invest $2,000 at 4% and $1,000 at 6%, or the entire amount at 5%. Which investment will yield him the larger income? (5)

(*c*) A man bought radio stock for $6,000. The value went down $12 a share when he could have bought 5 shares more for the same amount. What price did he pay for each share? (2)

Investment problems are given a wide range of values. Those that are given the highest value describe the most common practices of the average citizen in making investments, as in (*a*) above. The second example (*b*) describes a situation in which the average citizen would not find himself as often as he would in (*a*). The problems of this type that are given the least value, example (*c*), are expressed in the form of puzzles, the answers to which would have to be known before the problem could be stated.

6. *Lever.*

(*a*) A crowbar is 6 feet long. What weight could a man weighing 165 pounds raise, if the fulcrum were placed 8 inches from the other end? (5)

(*b*) A boy weighs 102 pounds, and he wishes to weigh his "buddy." He has access to a teeter board and a yard stick. Could he find the weight of his "buddy"? How would he proceed? (4)

None of the lever problems is given a value above five (5). While the principles underlying these problems are made use of frequently, the average citizen does not stop to calculate where he would place the fulcrum in order to raise a weight with a crowbar.

7. *Mensuration.*

(*a*) Some boys desire to lay off a circular race track on a vacant lot where they may practice for the annual meet. What radius must they use for a 150-yard track? (8)

(*b*) The City Hall of Philadelphia is said to cover the greatest area of any building in the United States. One-fifth of its width exceeds one-sixth of its length by 13 feet, and one-ninth of its length exceeds one-tenth of its width by 7 feet. Find its dimensions.

Mensuration problems describing situations involving dimension, area, volume—measurements that are in constant use—as example (*a*), are given high values as to importance. Others of this type have been converted into "catch" or nonsensical problems, as example (*b*). This latter class includes problems that no one would ever need to solve. In most instances, the answer is known before the problem is written. This class is ranked low in importance.

8. *Mixture.*

(*a*) A farmer has five Holstein cows each of which averages 6,000 pounds of milk within a certain length of time. The milk tests 3% butter-fat. How many Jerseys must he buy if each averages 5,000 pounds of milk testing 5% butter-fat, if he wishes to sell a mixture of the milk testing 4% butter-fat? (8)

(*b*) What quantity of alloy containing 15% gold must be mixed with one hundred ounces of an alloy containing 12% gold to make an alloy containing 13% gold? (6)

As shown by these representative examples, the mixture problems that make use of data relating to the common experience of people, such as percentage of butter-fat, reducing solutions of ammonia, medicines, etc., are ranked high in importance. Others, like example (*b*), that describe the more uncommon experiences, but make use of situations more or less cultural in nature, are ranked around five (5).

9. *Percentage.*

(*a*) Two boys have a chance to sell chocolate bars on a 30% commission. They wish to earn $7.80. How many bars must they sell at five cents each? (9)

(*b*) The Federal Government and the State Government each levy a tax of 20% and 10%, respectively, on all prize fights. The amount received by both from the Dempsey-Sharkey fight was $292,680. Find the gate receipts. (5)

(*c*) An angle is ½ its supplement. What percentage of the sum of the two angles is the angle itself? (2)

10. *Proportion.*

(*a*) Three men start in business and contribute capital in the ratio of 1: 2: 3. If at the end of the year the profit is $5,400, how much does each man receive? (9)

(*b*) The volume of a gas at 35 degrees centigrade and 790 millimeters pressure is 450 cubic centimeters. What is its volume at 90 degrees centigrade and 630 millimeters pressure? (4)

11. *Earning or Saving.*

(*a*) A newsboy makes $1.80 per week with 24 subscribers. How many subscribers must he find in order to earn $3.00 per week? (9)

(*b*) A father engaged his son to work on the condition that the son was to receive $2.00 for each day that he worked and was to forfeit $1.00 for each day that he was idle. At the end of 20 days, the son received $34.00. How many days did he work? (4)

A majority of the percentage, proportion, and saving problems that are ranked high in importance describe economic situations involving the common experiences of buying, selling, and the like, as shown by example (*a*). Those that are ranked low describe unusual situations, as shown in the last examples.

12. *Rate.*

(*a*) Cleveland and Columbus are 180 miles apart. A man in an automobile starts from Cleveland at noon and wishes to arrive at Columbus at six o'clock that evening. At what rate of speed must he travel on an average, if he makes no stops? (8)

(*b*) If Lindbergh's plane will fly at the rate of 120 miles per hour in a calm, what is the velocity of the wind when he can fly 375 miles with the wind in the same time that it takes him to fly 275 miles against it? (5)

(*c*) A man has an automobile which increases its speed from 5 miles per hour to 30 miles per hour in 10 seconds. What is his acceleration per second? (2)

Rate, or time-rate-distance, problems of the most common class, such as example (*a*), are placed rather high in the scale of importance. The introduction of means of rapid transportation makes it probable that a large part of our citizens will become involved in this kind of situation. It is considered very likely that only a few people would ever become concerned with finding acceleration as represented in example (*c*).

13. *Shares.*

(*a*) Twenty boy scouts were raising money for a two weeks' outing at a summer camp. They figured that they would need \$300. Five citizens each gave them \$10.00, and their leader promised to give them as much as they themselves would donate. One boy met with an accident so the others shared his expenses. How much did each boy have to give in order to raise the required amount? (8)

(*b*) A ditch laid through three towns for the purpose of drainage cost \$30,000. The towns agreed to pay according to the length of the ditch within their respective limits. Town *A* has the greatest length, *B* has $\frac{2}{3}$ as much as *A*, and *C* has $\frac{3}{4}$ as much as *A*. How much shall each pay? (5)

Most share problems describe economic situations in which the average citizen may find himself. All problems of this class are ranked high in importance. Example (*a*) is a case in point. Examples of this type, like example (*b*), however, describe unusual situations in which a small percentage of the people in general will be placed. This group is ranked considerably lower than those of the class described first.

14. *Work.*

(*a*) A farmer with a team can plow a field in 15 days. His son can plow the field with a tractor in 5 days. How long would it take them both working together? (8)

(*b*) A piece of work must be finished in 24 days. *A* can do it in 18 days and *B* in 30 days. Only one can work at it at a time. How long should *A* work so that *B* can finish it on time? (3)

The particular type of work problem, as shown in example (*a*), is given high value because a large number of people might have occasion to solve such a problem. Others, like (*b*), which portray unusual situations are ranked low in importance.

15. *Age.*

William's age is $\frac{1}{4}$ his age 20 years hence. How old is he now? (0)

16. *Angle Measure.*

The supplement of an angle is divided by its complement and the quotient is $2\frac{1}{2}$. What is the angle? (0)

17. *Digit.*

If one number is divided by the product of its two digits, the result is 3. Invert the order of the digits and the resulting number is in the ratio 7: 4 to the original number. What is the number? (0)

18. *Puzzle.*

(*a*) A shepherd on being asked how many sheep he had in his flock, said: "If I had as many more, half as many more, and seven sheep and a half, I should then have 500." How many sheep had he? (0)

(*b*) If the numerator of a certain fraction be doubled and the denominator increased by 5, the resulting fraction is $\frac{7}{8}$. If the numerator be decreased by 1, and the denominator increased by 7, the resulting fraction has the value $\frac{1}{3}$. What is the fraction? (0)

(*c*) I can think of two numbers whose sum is 23 and whose difference is 5. What are the numbers? (0)

(*d*) A fox is pursued by a greyhound, and has a start of 60 of her own leaps. The fox makes 3 leaps while the greyhound makes 2; but the latter in 3 leaps goes as far as the former in 7. How many leaps does each make before the greyhound overtakes the fox? (0)

All age, angle measure, digit, puzzle problems, etc., are placed at the bottom of the scale, as indicated by above examples.

SUMMARY

As shown by the above illustrative examples, the types of problems which rank highest in importance in the judgment of experienced and well-trained teachers of algebra are those that have to do with some kind of finance. Of all the problems that have been given a value of eight (8) or above, 50 per cent are expressions of some kind of economic situation. Thirty per cent of the problems which are evaluated around five (5) in importance are expressions of some kind of economic situation. The least important examples of these types are distributed among rates of travel, lever, percentage, relations of angle supplements, the relationship between volumes of a gas and the pressure exerted upon it, and the like. The remaining 25 per cent of the types that have problems ranking around eight (8) in importance include averages, indirect measurement, lever, mensuration, mixture, rate and work problems. The least important problems of the last-mentioned types are those that express acceleration of motion in automobiles, unnatural situation in purchasing stock on the market, and the like.

Note the types of problems having little importance. All age, angle measure, digit, clock, and puzzle problems have been evaluated at about zero (0) importance. It should be kept in mind that these problems may possess values of other kinds, but they are thought to have little application in life situations, when the average citizen is affected. "Application in life situations," it should be recalled, means "the necessity for solving or the occasion to solve."

These observed facts indicate that the greatest emphasis is now being placed on the most common experiences of the average citizen. It is interesting to note that of these most common experiences, the ones that deal with the actual welfare of the individual are given the highest rating. By welfare is meant, for example, the actual providing of funds, the construction of buildings, and the like.

The observations noted in the foregoing paragraph furnish a point of departure for the selection of problem material. If the pupil is to be shown the value of mathematics in everyday life, he must be given some material that is used and ought to be used in the world about him. This principle is a most important guide in the choice of mathematical subject matter.

CHAPTER V

ELEMENTS OF INTEREST IN ALGEBRA PROBLEMS

Interest as Estimated by the Teacher

In the statistical analysis of all data given in Chapter III, "Genuineness, Importance, Interest, and Difficulty in Algebra Problems," correlations between the various elements have been given. A rather high correlation is found there between genuineness, importance, and interest as estimated by experienced and trained teachers of algebra; that is, according to the opinions of these teachers, the chief element of interest in an algebra problem lies in its application to life.

Interest as Shown by the Pupil

The influence of economic pressure, which is evident in teachers' estimates, is not shown in the registration of interest in problems by pupils of junior high school age. The same high correlation between importance and interest is not found.

A list of illustrative examples will aid in the analysis of the factors in problems of varying degrees of interest to the pupil. These examples are chosen to represent all types of problems. Within each type, problems are chosen from various levels of interest. A comparison of these with the entire list studied shows them to be representative of the entire group.

Analysis of Representative Examples with Reference to Elements of Interest to the Pupil

It would be presumptuous to attempt to give a reason for the ratings given to each of the following examples by the pupils concerned in this study. However, an analysis of each with reference to its elements will aid in drawing some conclusions that may be of value to teachers of high school mathematics.

The number in parentheses at the end of each example indicates the interest value given it by the pupils.

1. *Age.*

(*a*) A teacher being asked how old he was responded: "At present I am twice as old as John. Ten years ago I was 17 years older." How old are the teacher and pupil? (7)

(*b*) A father promised his son a gift of $200 on the day that he would be twice as old as his son. At the time of the promise the son was 9 years old and the father 33. How old would the son be when he received his present? (5)

Example (*a*) obviously has the puzzle element. It also represents a situation in which the pupil or any child may find himself. It will be noted that this problem is given a high value as to interest. The second example, (*b*), also has a puzzle element. In addition to possessing the gift element, it describes a situation in which any child may see himself. This example is ranked fairly high also.

2. *Angle Measure.*

(*a*) The sum of the three angles of a triangle is 180 degrees. If the acute angle of a right triangle is three times the other, what is the size of each angle? (1)

3. *Clock.*

At what time between three and four o'clock are the hands of a watch opposite each other? (3)

Clock problems and those dealing with the measurement of angles seem to lack interest. Both describe situations that are static as far as pupil activity is concerned. The puzzle element is slight. There seems to be no particular *reason* for a pupil's solving either of the above examples.

4. *Average.*

(*a*) The average of a boy's grades in three subjects is 88. What must be his grade in a fourth subject in order that his average in all four subjects shall be 90? (7)

(*b*) The Giants have won twelve games and lost four. The Pirates have won ten and lost six. If, beginning with to-morrow's game the Giants should win and lose alternately each day, and the Pirates should win each day, how long would it be before the Giants and the Pirates would be tied? (3)

The *child* element seems to be the only one that is evident in example (*a*). This is a situation that is common to all pupils. Its interest value is comparatively high. Example (*b*) expresses adult activity. The situation involves the finding of averages that evidently the child is not interested in. It is reasonable to think that he would rather observe what is happening now and let the results be what they will.

5. *Formula.*

(*a*) Write a formula for finding the interest on $500 at 6% for 6 years, 9 months and 15 days. (6)

(*b*) A certain class treasurer always planned to leave $25 more in the treasury than there was when he took over the position. Write a formula to express this. Draw graph for formula. (2)

Most formula problems seem not to have a great deal of interest value. Six (6) is the maximum value given. The greater per cent of them express a static situation as example (*a*). Example (*b*) represents an unusual and an unreasonable situation. Its value (2) is observed to be low.

6. *Indirect Measurement.*

(*a*) James is 5 feet tall and Henry is 5 feet 6 inches. James stands 300 feet from the foot of a cliff, and Henry lies down with his feet against James's so that when he sights across James's head he can barely see the top of the cliff. How high is the cliff? (8)

7. *Lever.*

(*a*) A teeter is 16 feet long. Two boys, John weighing 85 pounds and George 75, are trying to determine how to place the fulcrum so that they will balance. Where will it be placed? (8)

(*b*) Two men are carrying a weight of 200 pounds placed on a pole between them. If the pole is 3 feet long, where should the load be placed so that the weaker of the two men will carry only $\frac{2}{3}$ as much as the stronger? (4)

8. *Mensuration.*

(*a*) A boy has a 48-inch strip of molding 1 inch wide with which he desires to frame a picture. His picture needs trimming and he would like to have it $\frac{2}{3}$ as wide as long. If he uses all the molding, into what lengths will he need to cut it if he miters the corners? (8)

(*b*) A railway company had to lay a track around a mountain. When laid, the track proved to be in the form of an exact semicircle and 11 miles long. How many miles would they have saved if they had tunneled through the mountain? (5)

(*c*) Two boys are to mow their grass tennis court which is 36 feet by 78 feet. How wide a strip must one boy cut around the court if he leaves just half for the other? (3)

Indirect measurement, lever, and mensuration problems seem to possess the same types of elements. For instance, take the (*a*) examples under the above types. First, there is activity expressed. Second, it is the kind of activity in which the pupil may participate. It expresses the activity of youth. Third, there is prevalent the play element. These examples have been placed high on the interest scale.

Note the (*b*) examples above. While the activity element is present it is adult activity. Example (*c*) under the heading *Mensuration* describes activity in which the pupil may find himself, but the element of work enters into the situation. It will be noted that in each of these cases the interest rating is low, especially is this true in the last case cited.

9. *Mixture.*

(*a*) A man buys two gallons of alcohol for his automobile. How many gallons of water must he add to have a 25% solution? (5)

(*b*) A grocer has 20 pounds of coffee worth 50¢ per pound. How many pounds of 35¢ coffee should he add to it to produce a mixture worth 40¢ per pound? (2)

Mixture problems are not generally given very high values in interest. In example (*a*), the highest in rank, a great many pupils might find an interest because they could see a *reason* for its solution. Someone closely connected with them, say a father, might *need* to solve such a problem. The situation, too, is suggestive of travel. Example (*b*) describes a situation in which only the children of merchants might become interested. Children in general would have very little *need* for the solution of such problems. The interest expressed in this last case is low.

10. *Number.*

(*a*) It: I am thinking of two numbers whose sum is 23 and whose difference is 5. What are the numbers? (8)

(*b*) If the numerator of a certain fraction be doubled and the denominator increased by 5, the resulting fraction equals $\frac{7}{8}$. If the numerator be decreased by 1 and the denominator increased by 7, the resulting fraction has the value $\frac{1}{3}$. What is the given fraction? (5)

(*c*) In a number of three digits, the tens' digit is twice the units' digit, and the hundreds' digit is twice the tens' digit. The sum of the digits is 28. What is the number? (4)

Number problems which may include fraction and digit problems vary according to the manner of stating each. As shown by example (*a*), those that are stated in the form of a definite challenge seem to create the most interest. Examples (*b*) and (*c*) possess possibly as much of the puzzle element as (*a*), but the challenge, "I am thinking," is lacking. Also, the last examples are probably more difficult of solution.

11. *Percentage.*

(*a*) Out of 25 true and false questions, Mary got 19 correct. The next week she answered correctly 23 out of 30. Which was the better score? (8)

(*b*) If $2,500 is invested at 6% and $35,000 at 5½%, how much must be invested at 8% to make the total income 7% of the total amount invested? (5)

(*c*) A retail store sold merchandise regularly at a profit of 30% on the cost. During a special sale the store offered all goods at a discount of 20% on the selling price. What was the store's profit on a sale of $10,400 at the special price? (2)

The percentage problems that receive the highest evaluation as to interest are of the class represented by example (*a*). Here, again, pupil interest or activity is involved. Examples (*b*) and (*c*) which drop down in interest value describe economic situations in which the child is usually not concerned. It is possible that example (*b*) is placed higher than example (*c*) because of its application to a greater number of people. There is more *reason* for its existence—it comes nearer affecting the happiness and comfort of the average child.

12. *Puzzle.*

(*a*) A shepherd on being asked how many sheep he had in his flock, said: "If I had as many more, half as many more, and 7½ sheep, I should then have 500." How many sheep had he? (8)

(*b*) If John should buy 20 marbles and Henry should lose 10, John would then have four times as many as Henry. How many has each? (5)

Problems stated in the form of puzzles are ranked high in interest. Example (*b*), although it expresses child activity, does not have that definite puzzle quality that example (*a*) has. There is a noticeable drop in interest in this last example.

13. *Time-Rate-Distance.*

(*a*) A boy has 3 free hours on a spring afternoon. His father agrees to drive him to the woods 5 miles distant in a car whose rate of speed is 25 miles per hour. The boy must walk back. If he walks 5 miles per hour, how long can he stay in the woods before starting back? (7)

(*b*) It takes a 250-foot train 2 minutes to completely pass over a bridge a mile long. What is the average rate of the train over the bridge? (4)

(*c*) A plane starts from New York to meet a ship sailing west. If the ship is 200 miles out and the rate of her speed is 20 miles per hour while that of the plane is 120, how far out will the ship be when the plane flies over her? (5)

All time-rate-distance problems are evaluated relatively high in interest. Of these, the ones that express child activity or that describe situations in which the child may find himself are given the highest value. This may be observed from the above examples.

14. *Saving.*

(*a*) A boy is putting his money in the school's saving bank with the idea of buying a scout suit when he has sufficient funds. If the suit costs $4.75 and he already has $1.25 to his credit, how long will it take him to save the amount if he deposits 50¢ weekly? (8)

Problems in saving as described above are evaluated high in interest. In this case the saving is to result in the satisfying of a childish desire.

15. *Work.*

(*a*) Two boys wish to finish picking a certain cotton patch one Saturday morning that they may have the afternoon for themselves. If there are 19 rows to pick, and one boy can pick 3 while the other picks 2, how many will each pick before they are through? (3)

All work problems regardless of the situation described are placed at the lower end of the scale of interest.

Summary

The above analysis of the representative examples throws light on the subject of interest in algebra problems. The elements of interest are perhaps too numerous, and many are too elusive for a complete tabulation of them. There are some elements, however, that are evident enough to warrant a definite statement concerning them.

1. As the data show, activity is one of the chief elements of interest to the pupil. Problems that express activity of youth in general are preferred. This is shown above by (*a*) under the

types: average, lever, mensuration, percentage, rate, and saving. If, however, adult activity is expressed in terms of racing, for instance, it is more acceptable.

Contrast this expressed interest with that for the last examples under the types: lever, mensuration, percentage, rate, and work. Problems in which a static condition is described, or in which activity is approached from the adult's point of view, or that are expressed in terms of work, all alike seem to create little enthusiasm.

2. Economic situations do not impress the pupil very much except where there is immediate gain to be realized by someone. There is a great deal of interest if the saving is to result in some immediate returns or in satisfying some childish desire. In economic situations in which adult activity is involved, the interest is less marked, ranging from five (5) to zero (0).

3. According to the data, pupils register pleasure in a definite challenge. Puzzle problems, stated as such, are liked especially well. To be told, however, that a given problem is a puzzle detracts decidedly from the interest. The same types of problems stated differently get different reactions from the same group of pupils. The following examples will illustrate this point:

1. It: I am thinking of two numbers whose sum is 23 and whose difference is 5. What are the numbers? (8)
2. A mathematics class was giving puzzles. John said, "I am thinking of two numbers. Their sum is 63, and one-half the larger equals two-thirds the smaller. What are the numbers? (5)
3. A teacher being asked his age responded: "At present I am twice as old as John. Ten years ago I was 17 years older." How old are teacher and pupil? (7)
4. A father promised his son a gift of $200 on the day that he would be twice as old as the son. If at the time of the promise the son was 9 years old and the father 33, how old would the son be when he received the present? (5)

Examples 1 and 2 are of about equal difficulty, and are of the same nature, but it seems that the statement "A mathematics class was giving puzzles" lessens the interest in the second. Similarly, Problem 3, worded in the form of a puzzle and of a challenge, has rather high interest value, whereas Problem 4, stated in the form of an ordinary problem, is rated comparatively low in interest.

4. Problems that have only a "semi-puzzle" element of interest do not create a great deal of enthusiasm. Clock, digit, and angle measure problems in particular are of this nature. While they do possess the puzzle element to a certain extent, they lack that definite challenge which is characteristic of the purely puzzle type. They not only lack the definite challenge, but also the element of activity and of immediate gain.

5. Reasonableness is probably a minor element of interest. Further evidence is needed to warrant a definite statement of this as a fact. Consider again the mixture problems on page 27. Example (*a*), with a value of five (5) in interest, lacks the elements considered above to raise its interest value very high, but there is some reason or need for its solution. There is a possibility of its occurrence to anyone owning a car. Example (*b*) with an interest value of two (2) belongs to a special class in which the pupil in general is not interested. Clock, digit, and angle measure problems, which lack all the elements considered above and which are low in interest, may be said also to lack reasonableness.

The findings, resulting from an analysis of problems with reference to elements of interest, should serve as a guide to the classroom teacher, for both the choice of problems and the method of presenting them. If the interest of the pupil is to be maintained, the teacher must overcome the temptation of choosing the problems that seem most interesting to her and of showing the pupil how to find the required answer. Instead she must proceed to set up a situation which seems real to the pupil; one in which he is interested. Then there should follow situation after situation, each appealing to the child as real and worthy of his attention. As is shown by the teachers' ratings of problems used in this chapter, there is a mass of material thought to be worthwhile in which the pupil finds little interest. It is the business of the teacher, then, to find some means of presenting this material so that it, too, will become interesting.

CHAPTER VI

GENERAL SUMMARY

The foregoing chapters are concerned primarily with determining what problem material is considered most important in leading the child to discover his relationships to his environment, and can be used to advantage in preparing him to participate more efficiently as a citizen in the life of the community.

It is the opinion of a large, representative group of teachers of mathematics that the problems with which the average citizen is most concerned are those that have to do with economic situations; that is, the providing of funds for maintenance, buying and selling, investments; costs, such as, shipping, mailing, sending telegrams, riding in taxis, and the like. Other important problems are based on mensuration, such as, building, construction work, determining distances or heights indirectly, traveling, marketing, mixtures, and the like. Stated in more general terms, the problems that deal with any kind of material advancement are given first choice as to importance.

Age, digit, clock, and all kinds of puzzle problems, which formerly constituted such a large part of problem material, are placed at the bottom of the scale of importance by the teachers. This discredits somewhat the old idea that "It makes no difference what is taught just so it's hard."

There is high correlation between the degree of importance and the amount of interest that the teacher thinks the pupil will show in each type of problem. If an example describes a situation which would result in the betterment of a large number of people, the teacher is inclined to think that this is a sufficient basis for the pupil's liking it.

The pupil expresses a somewhat different opinion with respect to interest and importance. His interest in a problem generally does not depend upon its importance, as defined by the teacher. There is likewise practically no correlation between interest as estimated by the teacher and that expressed by the pupil.

Too many difficult problems given at one time tend to lessen the interest of the pupil. The pupils, however, express a liking for difficult problems of certain types if only a few are introduced at a time and if these are well distributed among the ones of less difficulty. These facts were pointed out under Difficulty and Pupil Interest in Chapter IV. As is shown there, the negative correlation between interest and difficulty is not only lower, and the mean interest value higher, when the problems are less difficult, but the problem that ranked highest in interest was the most difficult.

Teachers overestimate their ability to create interest in problem-solving. Seventy-five per cent of the problems used in this study, that had been taught to one group of pupils, were rated higher in interest by the teacher who had taught them than by the pupils themselves.

As is pointed out in Chapter IV, the pupil is most interested in problems that express activity; that describe situations in which he may participate. A problem stated in terms of child activity receives higher evaluation by the pupil than the same problem expressed in terms of adult activity.

Likewise, as Chapter IV shows, the pupil likes a problem better when it is stated so as to give a definite challenge. It makes little difference whether this challenge involves a puzzle or a genuine problem.

The findings summarized above make it evident that the classroom teacher is confronted with a definite problem. He is given a mass of material that is important in that it has direct application in the lives of the average citizen. In general, he is convinced that the pupils should be interested in such material. The pupil, however, is not necessarily interested in problems because they are important. For instance, static economic situations involving the provision of food, clothing, and shelter do not arouse his enthusiasm.

The problem of the teacher in this connection is one of determining some means of presenting this valuable material so that it will appeal to the pupil and, in addition, will give him a method of attack that will enable him to discover his own solutions.

By taking cognizance of the relationships found to exist between the factors, *genuineness, importance, interest,* and *dif-*

ficulty, and by noting the elements of importance and interest to the pupil, the teacher is enabled not only to select problems and problem material more wisely, but is also able to present the work more interestingly and effectively.

APPENDIX A

SALIENT POINTS IN METHODS OF TEACHING PROBLEMS IN ALGEBRA

In the course of this investigation it was necessary to examine all of the literature regarding problem-solving in algebra. The study of this material, together with the findings of this dissertation, have given the author a new point of view concerning methods of teaching problem-solving in algebra, which is summarized below.

It must be understood, however, that these conclusions have not been experimentally verified except in the respects mentioned in the body of the thesis. They are given here in order that this study may have as much practical value as possible.

1. A Brief Survey of the Literature on Problem-Solving

The teaching of algebra by inexperienced teachers results too often in the mere manipulation of symbols with a minimum of time and effort given to one of the most valuable phases of the subject—that of problem-solving. The blame for this state of affairs does not lie wholly with the beginner, but largely with those who through long years of experience have developed methods whereby they are able to make the subject alive and rich in content, and have not yet made this knowledge generally accessible to others.

A brief survey of the literature on the teaching of mathematics reveals a scarcity of material on problem-solving. Evidently too little attention has been given to this phase of algebra teaching. Articles of a general nature are plentiful, but suggestions as to the methods of teaching are often vague and give little help to the inexperienced teacher.

Nunn's *The Teaching of Algebra Including Trigonometry* [1] furnishes a rich source of material, together with valuable sug-

[1] Nunn, T. P., *The Teaching of Algebra Including Trigonometry*. Longmans, Green & Co., 1913.

gestions for teaching. Though the book was written some time ago, Nunn's idea of the correlation of branches of mathematics and of the notion of functional relations is modern even in the light of more recent developments. Again, in *The Psychology of Algebra,*[2] Thorndike has laid down some vital concepts. Both of these works are somewhat difficult and the young and inexperienced teacher would perhaps need direction in reading them in order to derive the greatest benefit therefrom.

Various short articles by McClymonds,[3] Lytle,[4] Dines,[5] Booker,[6] and Nyberg[7] give interesting discussions as to the value of functionality and the use of the graph as an easy and effective means of conveying this concept. Other articles on the graph, in particular, indicating how it may be used as a means of picturing relationships are given by Christofferson[8] and by David Eugene Smith[9] in one of his short classics, "The Lesson of Dependence."

There are available only a few books and articles that give specific information or hints on how various types of problems can be taught. Rudman[10] gives some valuable suggestions on the use of the analytic method in general and makes specific application of this method in a few instances. Ligda in *Teaching Elementary Algebra*[11] makes the function concept the central theme, but the method that he uses in problem-solving is novel, and not one that the beginner could commonly employ. Nyberg[12] suggests the use of the rule and chart for certain types of problems. This method if used by the average teacher would tend to become a mechanical device for securing answers, thereby

[2] Thorndike, E. L., *The Psychology of Algebra.* The Macmillan Co., 1923.

[3] McClymonds, J. W., *Teaching Problem-Solving in Algebra II.*

[4] Lytle, E. B., "Suggestions on Teaching Elementary Algebra." *Education,* Vol. 31, pp. 175-78, November 1910.

[5] Dines, L. L., "Development of Function Concept." *School Science and Mathematics,* February 1919.

[6] Booker, E. E., "Teaching Function Concept in First Year Algebra." *The Mathematics Teacher,* February 1926.

[7] Nyberg, J. A., "Prose Problems of Algebra." *School Science and Mathematics,* December 1920.

[8] Christofferson, H. C., "The Graph as a Means of Picturing Relationships." *The Mathematics Teacher,* April 1928.

[9] Smith, David Eugene, "The Lesson of Dependence." *The Mathematics Teacher,* April 1928.

[10] Rudman, Barnet, "Teaching the Verbal Problem in Intermediate Algebra," *The Mathematics Teacher,* February, 1929.

[11] Ligda, Paul, *The Teaching of Elementary Algebra.* Houghton Mifflin Company, 1925.

[12] Nyberg, J. A., "Prose Problems in Algebra," *School Science and Mathematics,* December, 1910.

defeating the purpose in teaching of problems; viz., showing mutual dependence and developing ability to generalize.

2. The Problem of Method

Important Problems and Pupil Interest. As is shown in Chapter III, the problems that are considered most important by a large group of teachers of algebra are those that have to do with economic situations involving adult interests and those that express a static situation.

It is also shown (in the conclusion of Chapter IV) that none of these factors appeal to the interest and imagination of the child. On the other hand, pupils indicate an interest in challenges and puzzles, and problems that describe action or situations in which they may participate.

Purpose. The purpose of this section is the development in a brief but concise manner of some methods and procedures in problem-solving. These methods and procedures, drawn largely from personal experience, have been found to secure a great deal of interest and at the same time to retain the desired rigorousness.

3. Abilities in Problem-Solving [13]

Abilities in Algebra, in general, have been treated extensively in *The Teaching of Junior High School Mathematics,*[14] by Smith and Reeve, but it might serve to establish the desired point of view if a few abilities in problem-solving are given here as a basis for further procedure.

1. *Ability to select essential data given in an algebra problem.* This means that the pupil should be able to choose from a group of data the essential material, and disregard irrelevant facts, for the purpose of setting up equations for the solution of the problem.
2. *Ability to form equations.* This means that the pupil should be able to express in short-hand the equality existing between essential data in a problem.
3. *Ability to express equations in general terms.* This means that after equations have been formed from specific data, the pupil should be able to express them in general terms, thereby showing that a law of relationship can be expressed for almost any kind of situation.
4. *Ability to construct graphs.* This means that after equations for specific facts have been generalized, the pupil should be able to represent them graphically.

[13] This list of abilities is based on "Abilities in Algebra" given by Smith and Reeve, *The Teaching of Junior High School Mathematics*, pp. 67 ff. Ginn and Company, 1927.

[14] *The Teaching of Junior High School Mathematics.* Ginn and Company, 1927.

5. *Ability to interpret graphs.* This means that the pupil should be able to see at a glance how one quantity varies with another.
6. *Ability to estimate rates from graphs.* By drawing tangents to a curve at various points the rates of change or rapidity of growth in the dependent variable can be compared for different values of the independent variable.
7. *The ability to assign specific problems to types.* This means that the pupil should be able to recognize a specific problem as belonging to rate, time, distance, etc., types.
8. *Ability to discern equalities which usually exist in certain types of problems.* This means that pupils should see readily that in time–distance–rate problems, for example, time, distances, and rates are usually equated.

4. Procedure in General

A consideration of the above abilities will lead one to anticipate the general method in the teaching of algebra problems that follows:

Division of Problems as to Types. It is not advised that all problems be separated into types and that each type be taught as a unit distinct and apart from other units of work. The instructor, however, should make it a point to see that after the pupil has met with a few simple cases of a certain type, he should be able to assign all such problems to a definite category, and later be led to discover a general method of attack on such types. Experience has proved that by following this procedure pupils develop a sense of security much earlier and show remarkable interest in using this newly acquired knowledge in similar cases.

Problem Analysis. A majority of pupils approach the verbal problem with uncertainty and confusion of mind. They are given explanations by the teacher, but very often they lack guidance in original work. They need some method of arriving at the origin of the logical statements made by the teacher in the explanation of the solutions of the problems. If the pupil can be taught to ask himself proper questions concerning the facts given in a problem, he is well along the way toward being able to think his way through it. Only in this way can he find the constituent parts and discover the relationships between them.

The suggestions in the preceding paragraph point to an analytic method of solution. The first step in the mastery of this method is to develop in the pupil the ability to recognize a problem as belonging to a specific type. The second step is to give the pupil a general method of attack in the solution of problems of each type. Within each type of problem there are certain elements that may be equated. For instance, in time–rate–distance problems, equal amounts of time, equal rates, or equal distances are involved. The third step is concerned with determining what is required and the fourth step is concerned with taking note of either data given or to be supplied.

As a basis of analysis, the following questions are readily recalled by the pupil, and are referred to hereafter as *steps*.

1. What type of problem is this?
2. Equality usually exists between what elements in this type of problem?
3. What is to be found?
4. What do I have to find it with?

If the pupil can answer the first question, he feels more confident in going to the next. Answering the second question will present little difficulty if the general method for solving each type of problem has been well established.

Graphical Analysis. Not all types of problems lend themselves equally well to graphical analysis. The construction of graphs for some of these would probably be a waste of time. A consideration, however, of the probability or advisability of a graph for all types will aid in developing the habit of looking for relationships. The graph, table, and formula serve as some of the best means of showing relationship and should be introduced early. The function concept, as such, would probably better be deferred until later. The graph is not only the best method of showing relationships, but it makes an appeal to the eye. If used wisely, it is one of the best means of increasing interest.

5. Detailed Analysis of Type Problems

Illustrative examples such as are found in textbooks are desirable for use as models. These will help the pupil to develop a neat, concise form for writing solutions. But this alone is not sufficient for the pupil's needs. Some kind of method should be given him whereby he will be able to discover more readily his own solution. A consideration of representative examples from several types will serve to clarify the general method described above in "Problem Analysis."

(*a*) *Age Problems.*

Consider the following problem as a representative of this type:

> A father promised his son a gift of $500 on the day that the son would be one-half as old as he. Their ages now are 35 and 10, respectively. How many years will the son have to wait for his gift?

Following the procedure laid down in problem analysis:

(1) "What type of problem is this?" The type of problem is "age." Even though the type is very evident, the pupil should form the habit of at least *thinking* the type. This should lead him directly to question 2.

(2) "Equality usually exists between what elements in this type of problem?" In this type, equality usually exists between sums or differences of ages, or known or discernible ratios between ages. The pupil should be taught to recognize this at once, and either to think or to write:

$$\frac{A\text{'s age}}{B\text{'s age}} = \frac{?}{?}$$

$$\text{or} \quad A\text{'s age} \pm B\text{'s age} = ?$$

Then, without stopping to answer these questions, he should go directly to the next two:

(3) "What do I want to find?" and

(4) "What do I have to find it with?"

These questions necessitate a recalling of the data given in the problem from which the pupil may write:

$$\begin{aligned} x &= \text{the number of years to wait.} \\ \text{But} \quad 35 \text{ years} &= \text{the age of the father now.} \\ 10 \text{ years} &= \text{the age of the son now.} \\ \text{and} \quad \tfrac{1}{2} &= \text{the ratio of their ages after } x \text{ years.} \end{aligned}$$

Since the boy is interested in x years, he is interested in the ages at the expiration of x years. Hence, he would add:

$$\begin{aligned} 35 + x &= \text{the age of father } x \text{ years hence.} \\ \text{and} \quad 10 + x &= \text{the age of son } x \text{ years hence.} \\ \therefore \quad \frac{10 + x}{35 + x} &= \frac{1}{2} \end{aligned}$$

Generalization should be made concerning this last equation by pointing out that it can be changed to a formula that will express this relationship between the ages of any boy and his father. This is done by changing the specific ages, 10 and 35, to *B*, boy's age, and *F*, father's age. The formula becomes:

$$\frac{B+x}{F+x}=\frac{1}{2}.$$

Then, by substituting the present ages of any boy and his father, the number of years can be found in which this boy will be just half as old as his father.

(*b*) *Lever Problems.*

Lever problems are usually not complicated, hence a simple explanation of the law, force equals weight times distance from fulcrum, is sufficient. This law may be expressed algebraically by the equation,

$$F = W \times d.$$

If a state of equilibrium exists, the forces (F) are equal and the equation becomes:

$$W_1\, d_1 = W_2\, d_2.$$

The usual division of lever problems into classes as such, viz., first, second, and third, is not necessary for high school algebra work, but the principles involved in each class should be explained and the situation described or illustrated. Especially is this true for the cases in which the fulcrum is between two weights, and in which the fulcrum is at one end and the weight is between the fulcrum and an upward force. The first case is very simple, but the last would perhaps give some trouble. The distribution of the weight of an automobile on the two abutments of a bridge is a case in point. Take the problem:

> An automobile weighing 2,400 pounds is driven across a bridge with a span of 100 feet. How much weight is exerted on each abutment, *A* and *B*, when the automobile is 30 feet from abutment *A*, assuming that the weight of the automobile is concentrated at one point?

Here it must be shown that each abutment must act as fulcrum in turn and that the other exerts an upward force acting through the entire length of the span of the bridge. This upward force is equal to the force exerted by the weight of the car acting through its distance from the opposite abutment.

Going through the four steps in the analysis laid down earlier: (1) the type is easily established; (2) equality usually exists between weight 1 times its distance and weight 2 times its distance, both distances being measured from the fulcrum. This equality may be expressed:

$$W_1 d_1 = W_2 d_2.$$

(3) What is to be found requires only a glance over the problem to determine weight 1 (W_1) on, say, abutment A first, and second on abutment B. Step 4, answering the question, "What do I have to find it with?" may require a careful re-reading of the problem. All the known data and the general equation just formed are then related.

The pupil then is ready to begin the deductive process, as:

W_1 = the weight supported by abutment A.
d_1 = 100 feet distance abutment A is from fulcrum B.
W_2 = 2,400 pounds, the weight of automobile.
and d_2 = 70 feet distance of W from fulcrum B.
$\therefore\ 100W_1 = 70 \times 2{,}400$ pounds.

In a similar manner, replacing A with B and 70 feet with 30 feet, $100W_1 = 30 \times 2{,}400$ pounds expresses the weight on abutment B.

To stop the problem here would mean the loss of an opportunity to show more clearly and forcibly the idea of relationship. The greatest challenge that could be given is the question, "What happens as the car moves across the bridge?" The pupil can be led to discover that the general formula,

$$W_1 d_1 = W_2 d_2$$

becomes

$$100w_1 = 2{,}400\ d_2.$$

d_2 *and* W_1 being constants. Hence, for each value of d between 0 and 100, a different weight, W_1 can be found varying from 0 pounds to 2,400 pounds.

This being a linear equation, it can be quickly represented by a graph and the approximate values read from this. Additional interest can be obtained by plotting graphs of both the equations for the weight on A and B by holding d_1 as the distance of the automobile from B and letting W_1 represent weights for both A and B, respectively, noting various points of interest.

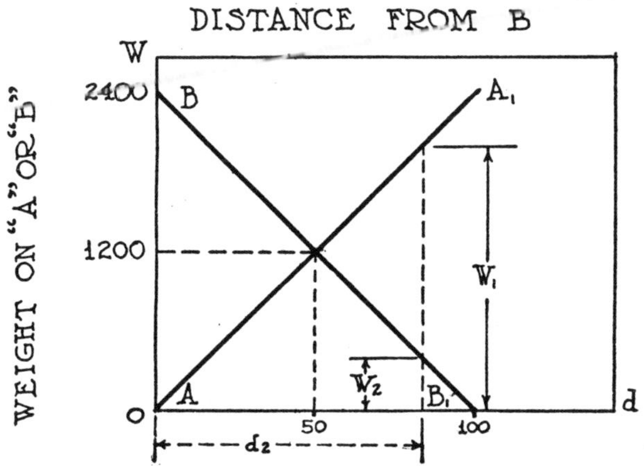

DIAGRAM III

DEPENDENCE OF FORCE UPON WEIGHT TIMES DISTANCE FROM FULCRUM

From Diagram III it can be seen that the point of intersection of the two lines represents half the length of the bridge and half the weight of the automobile. By drawing lines perpendicular to the d axis, the weight (w_1) on A varies directly, and the weight (w_2) on B varies inversely as the distance d_1 from B.

(*c*) *Mensuration Problems.*

Mensuration problems may be made to include a number of sub-classes or types dealing with all kinds of measurements, such as, perimeters or rectangles, circumferences, areas, volumes, indirect measurement, including problems of proportion, and trigonometric measurements of the right triangle. The teaching of problems of this type presents little difficulty. Hence, an indication of how the meaning of some of these problems may be extended and of how more interest may be injected is all that is deemed necessary.

The method of solution in most instances involves the use of formulas. This makes it possible to show relationships and at the same time increase interest. The question, "What will happen if ——?" gives the challenge and arouses the curiosity of the pupil. This is particularly true in the case of the relationship between the radius and the circumference or the radius and the area of a circle. For instance, consider the problem:

The track team wishes to lay off a ¼-mile circular track on the athletic field. What radius should they use?

The solution of this particular problem is obtained by setting up the formula

$$C = 2\pi r$$

and filling in the data given in the problem. This gives the equation

$$440 \text{ yards} = 2\pi r$$

and the radius expressed in yards is found. But the problem should not be stopped there. The question of extending the track should be brought out. What would the radius have to be to make the track ½ mile long? 1 mile? This discussion should lead to the relationship in general between the radius and the circumference, i.e., what is the effect on the circumference of a circle if the radius be doubled, tripled, quadrupled?

Graph and note that the ratio is constant, i.e., that the graph is a straight line. This leads to the discovery of π. The discussion can then be turned easily to a consideration of the area of the playground enclosed by the track as a function of the radius. What would be the effect upon the area as the radius is changed? This leads to a consideration of the formula,

$$A = \pi r^2.$$

The interest can be heightened at this point by plotting the graphs of these two formulas on the same axis. Care should be taken that the units on the functional axis (the vertical axis), while numerically equal, represent linear units in one case and square units in the other.

It can be seen from the graph that the area of the circle increases very slowly at first, being less numerically than the circumference until it reaches the point where $r = 2$, then it passes its opponent and rises rapidly.

It can also be observed from the graph that the rate of change in C is gradual and constant,

$$\frac{\Delta c_1}{\Delta r_1} = \frac{\Delta c_2}{\Delta r_2}.$$

By drawing tangent T_1T_1, which represents the rate of change in A, to the curve of $A = \pi r^2$, it may be observed that at some point P the area is changing at the same rate as the circumference. Then by drawing another tangent, T_2T_2, at the point P_1

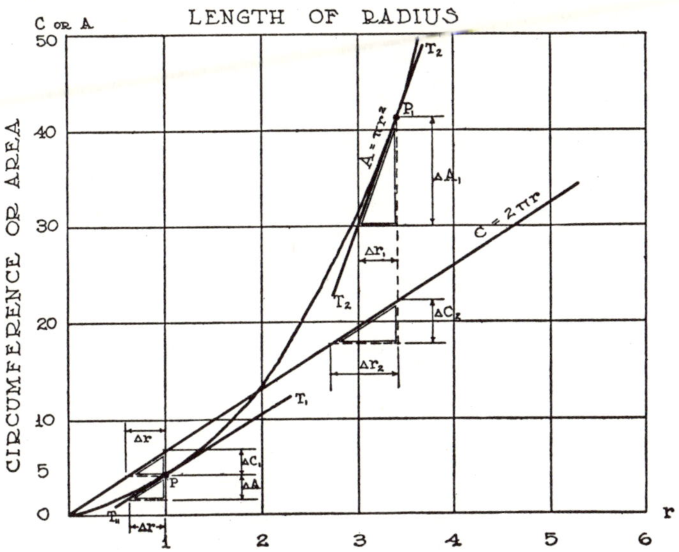

DIAGRAM IV

GRAPH SHOWING COMPARISON OF AREA WITH CIRCUMFERENCE OF CIRCLE

for another value of r, the rate of increase in the area is very rapid, being represented by the slope of the tangent.

The introduction of the tangent and of the increment as means of showing rates of change should in most cases be postponed until the last two years of high school. Here it may be employed not only to increase interest, but as a basis for elementary calculus.

The same procedure as that suggested by Diagram IV for the graph may be followed and comparisons made for the area and volume of the sphere or of the cube.

One other case that might be discussed with profit is that of a maximum area enclosed when given a definite perimeter. Take for example:

> A man has a roll of fencing containing 300 feet of net wire. He wants to enclose the largest possible rectangular chicken yard. What dimensions should he use?

Here again the pupil sets up a formula,

$$A = lw$$

in answer to step 2, What equality usually exists——? From 3, What is to be found? l and w are readily discovered. But the last step, 4, What have I to find it with? brings difficulty. It is discovered that there is not sufficient data with which to obtain a definite result. There is only one specific quantity given and one equation with three unknown quantities. The pupil's problem, therefore, becomes one of eliminating all the unknowns in the formula with what material he has given.

$$A = lw$$

He can discover l in terms of 300 and w; thereby reducing the number of unknown quantities in the equation to two. This leads to:

$$l = 150 - w \text{ (half the perimeter minus width)}$$

$$\text{whence} \quad A = w(150 - w).$$

The pupil should be led to discover at this point that the solution can be arrived at only through approximations and that the closest approximation comes through careful graphing. Hence, by compiling a table of values for A and W and with the graph drawn as smoothly as possible, he has:

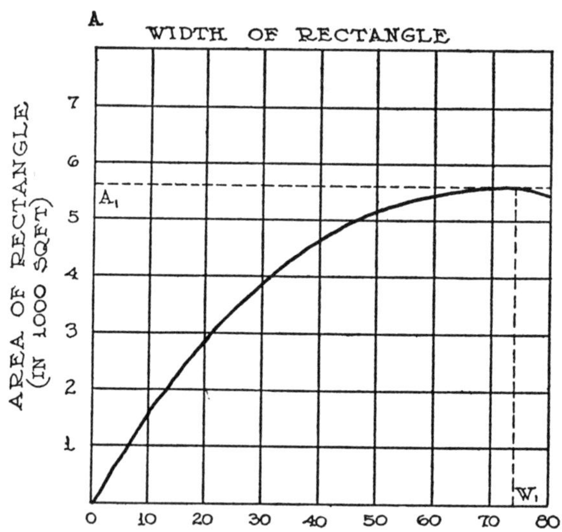

W	0	----------------	50	75	---
A	0	----------------	5000	5625	---

DIAGRAM V

RELATION OF AREA OF A RECTANGLE TO DIMENSIONS, GIVEN A FIXED PERIMETER

It will be observed that there is a highest point on the curve. By drawing a horizontal line across the top of the curve, a value on the vertical axis will give the approximate area (A_1). By dropping a perpendicular from this highest point to the W axis, the approximate width (W_1), which gives a maximum area, is found. Tests should be made by substituting values on each side of W, to determine whether or not this is the maximum A.

(*d*) *Mixture.*

The attack on problems dealing with solutions, alloys, and other mixtures is very much the same. One illustration will serve to give a general method for solving that will be applicable to all. Take the following example:

How many gallons of water must a farmer add to 40 gallons of milk testing 5% butter-fat if he wishes to sell milk testing 4% butter-fat?

Proceeding at once to the analysis, step 1 is determined and (2) What equality usually exists? is found to be that between percentages or hundredths. (3) What do I want to find? leads to a discovery of x gallons of water to make a $\frac{4}{100}$ solution. This ratio should suggest the general ratio, $\frac{\text{No. of gal. of butter-fat}}{\text{No. of gal. of mixture}}$. The answer to (4) What do I have to find it with? leads to a discovery of 40 gallons of milk and 5% of 40, or 2 gallons of butter-fat. The pupil should have discovered earlier, that, to obtain a per cent, units of measure must be divided by like units. This results in the discovery that in order to get a $\frac{4}{100}$ solution from a $\frac{2}{40}$ solution, the water must be added to the milk. Hence, if

$$x = \text{the no. of gal. of water to be added.}$$

$$2x = \text{the no. of gal. of butter-fat.}$$

and $$40 = \text{the no. of gal. of milk.}$$

Then $$x40 = \text{the no. of gal. of mixture.}$$

$$\therefore \quad \frac{2}{x+40} = \frac{4}{100} \text{ or } \frac{1}{25}.$$

(*e*) *Proportion.*

Although the method of solution of problems of this type is comparatively simple, a large percentage of high school graduates enter college classes in chemistry and physics without being able to form proportions correctly.

Problems of this type can best be approached through the use of the formula $A = RN$, which expresses relations between total amounts (A) and certain rates (R) and quantities or numbers (N). Here, equality exists between amounts, between numbers, or between rates. To illustrate, take the simple case:

> Two families together purchased $6\frac{1}{4}$ dozen young fruit trees for $21.75. One family took 45 trees. How much should that family pay?

Following the general procedure suggested earlier in the discussion, the *type* of problem is not so evident. This may not be discovered by the pupil unless he has developed the notion that total costs and total quantities are, respectively, proportional to partial costs and partial quantities and expressed in the same units of measure, respectively. The second question, Equality usually exists between what elements? leads to the discovery that the same rate, or cost per tree, is paid by both families. Hence, the general formulas for both total cost and partial cost may be set up:

$$C = RN$$
$$c = Rn$$

from which the R may may be eliminated by dividing equals by equals, obtaining,

$$\frac{C}{c} = \frac{N}{n}.$$

The third and fourth steps, What do I want to find? and What have I to find it with? result in substitutions in the last formula, giving:

$$\frac{\$21.75}{C} = \frac{75}{45}, \text{ not } \frac{6\frac{1}{4}}{45}.$$

The usual form may then be followed, as:

Let C = the amount paid for the 45 trees.
$21.75 = the amount paid for all trees.
45 = the number of trees one family bought.
75 = the number of trees both families bought.

$$\therefore \quad \frac{\$21.75}{C} = \frac{75}{45}.$$

$$\text{or} \quad \frac{C}{\$21.75} = \frac{45}{75}.$$

(*f*) *Percentage Problems.*

The solution of a large number of percentage problems belongs to the field of arithmetic. However, a portion of that class of percentage problems that describes economic situations belongs to algebra by virtue of the fact that these problems can be expressed as formulas. Especially is this true of interest problems. It is here that the most helpful suggestions can be offered.

According to the analysis given in Chapter IV, this type of problem, while one of the most important, is one of the least interesting to the pupil. In order that interest may be heightened, some kind of motion might be introduced. Rather than attempt this through changes in the wording of the problem so as to make it sound childish—thereby sacrificing its rigorousness—the problem itself should be brought into action. The pupil should see it as something growing and expanding. The best means of doing this is found in the employment of the graph.

While the simple interest is not used except for very short periods of time, it would be well to let the pupil see the effect of different rates of interest. Take, for instane, the rates of 4 per cent, 6 per cent, and 8 per cent, and plot the graphs of simple interest on $100. Use the same axis and show how the lines spread as the time increases:

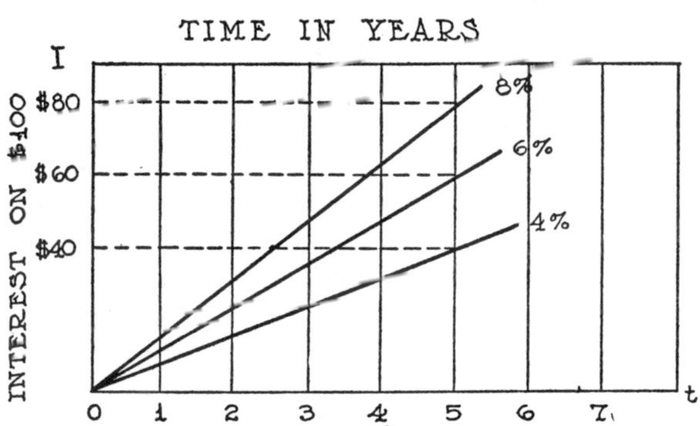

DIAGRAM VI

DIFFERENT SIMPLE INTEREST RATES

Now erect a perpendicular to the t axis at any point, and, from the points of intersection of this with the interest lines, drop perpendiculars to the I axis. Note the ratios of the interest rates themselves. This is another way of indicating that while the interest accumulates more rapidly at one rate than at another, as shown by the differences in the slopes of the lines, the rate of increase within each does not change. Then for the same rate of interest construct graphs on the same axes for simple interest and compound interest and compare these at various intervals of time.

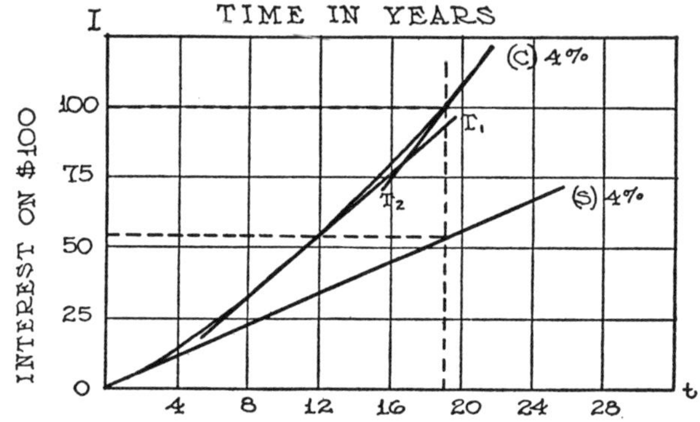

DIAGRAM VII

COMPARISON OF COMPOUND INTEREST CURVE WITH SIMPLE INTEREST AT THE SAME RATE

This can be made more impressive by drawing tangents T_1 and T_2 to the compound interest curve (C) at various intervals and comparing their slopes with the slope of the simple interest line (S). This will indicate how the rate of accumulation of interest increases upon compounding the interest.

In like manner, make comparisons between different rates of interest compounded at the same interval and between the same rate compounded at the different intervals of time. In Diagram VIII, Figure 1 shows a comparison of 4 per cent and 6 per cent compounded annually, and Figure 2 shows a comparison of 4 per cent compounded quarterly, semiannually, and annually.

Note the differences in the slopes of the tangents at different stated intervals of time. These graphs will show more forcibly the effect of increases in interest rates if computed on $100 or, better, $1,000.

(*g*) *Time–Rate–Distance.*

The method of attack on time–rate–distance problems can be made rather definite. (1) There is no difficulty in recognizing

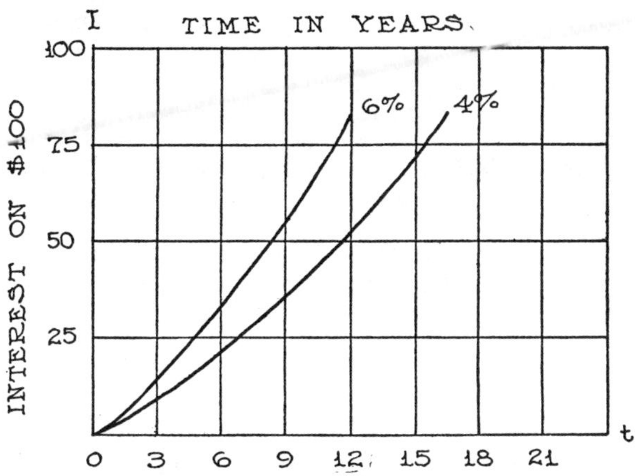

FIGURE 1

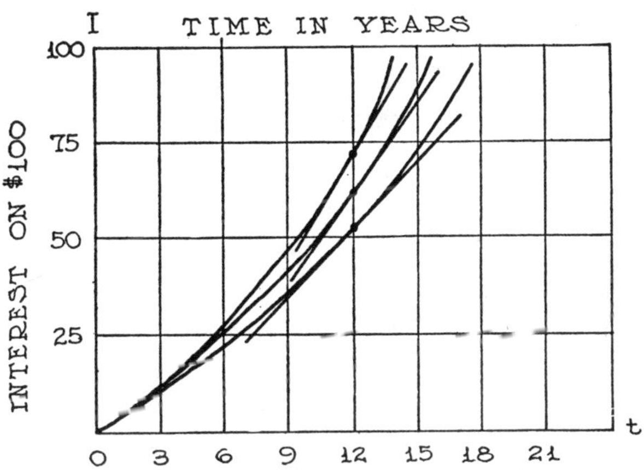

FIGURE 2

DIAGRAM VIII

FIGURE 1 SHOWS COMPARISON OF 4 PER CENT COMPOUNDED INTEREST CURVE WITH THAT OF 6 PER CENT, INTEREST COMPOUNDED ANNUALLY.

FIGURE 2 SHOWS CONTRAST OF 4 PER CENT INTEREST CURVE, COMPOUNDED ANNUALLY, SEMIANNUALLY, AND QUARTERLY.

the type. (2) Equality exists rather consistently between either time or distance elements. Sometimes it is convenient to equate rates. The general formula for this type is:

$$D = r \times t.$$

The habit of noting whether or not the same distance has been covered, the same amount of time has been consumed, or whether or not rates are compared should be formed. Take a specific case:

> Starting out from camp on the banks of a stream, James rowed 1 mile upstream at a steady rate in 1 hour. With approximately the same effort, he rowed back to camp in $\frac{3}{4}$ hour. What was his rate of rowing in still water, and the rate of the stream?

That equality exists between distances is quite evident, and, since there are two rates, there must be two equations, hence the equations:

$$r_1 t_1 = d$$

and

$$r_2 t_2 = d.$$

But the difficulty arises when substitutions are made with step 3 What I have. The quantities t_1, t_2, and d are known, but it must be established that the rate upstream is always the rate of rowing less the rate of the current, and the rate downstream is the sum of the two rates. In this case both are unknown, and equations become:

$$1(r_1 - r_2) = 1$$

and

$$\tfrac{3}{4}(r_1 + r_2) = 1.$$

Thus the pupil may write:

If r_1 = rate of rowing in still water
r_2 = rate of stream
1 hr. = time upstream
$\frac{3}{4}$ hr. = time downstream
then $r_1 - r_2$ = rate upstream
and $r_1 + r_2$ = rate downstream.
Hence, $1(r_1 - r_2) = 1$, the distance upstream
and $\frac{3}{4}(r_1 + r_2) = 1$, the same distance downstream.

The idea of dependence can be shown more easily and will arouse greater interest if the two equations $r_1t_1 = d$ and $r_2t_2 = d$ are graphed simultaneously on the same axis, that is, complete each line up to a certain time, then successively up to the next unit of time, and let the pupil observe the growth of each over each unit of time. To illustrate, take the problem:

> Two families plan to tour the West together. James' family was ready to start first, and it was decided that they start ahead and travel at

about 25 miles per hour, and, when John's family was ready, they would travel at about 35 miles per hour and overtake them. If James had been gone 3 hours before John started, how long would it take John to overtake him, and how far would they have traveled?

Here equality exists between distances traveled, and the equation for each car is:

$$d = r_1 t_1$$

and $$d = r_2 t_2.$$

Now, since the watches of the two groups are together, the time of travel can be expressed in terms of the time traveled by either of the groups, i.e.,

$$t_2 = t_1 - 3$$

since the second group starts 3 hours late. The two equations then become:

$d = r_1 t_1$ for the first automobile

and $d = r_2(t_1 - 3)$.

These equations can be simplified by step —"What have I to find it with?" which is the r_1 and r_2. Hence:

$$d = 25\ t_1$$

and $$d = 35\ (t_1 - 3).$$

Plot the first graph up to the time of starting of the second automobile—

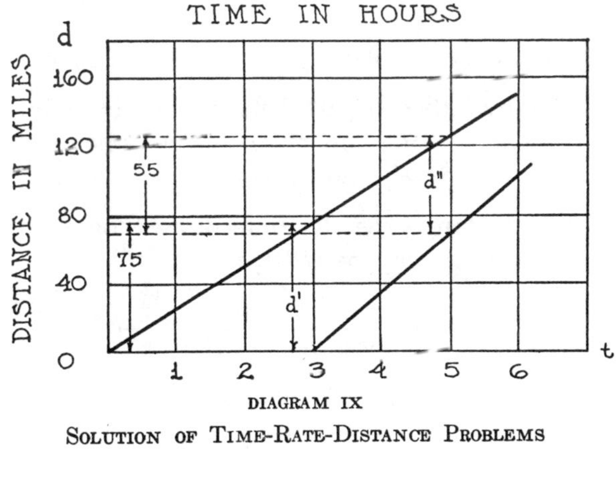

DIAGRAM IX

SOLUTION OF TIME-RATE-DISTANCE PROBLEMS

and note, on the line $t = 3$, the distance d' that James is ahead of John at the time of starting of the latter. Extend the lines to cover the next two hours and note the distance d'', and so on, for d''', d'''', comparing the difference between these distances each time. Continue until the lines intersect, reducing the time interval after each observation, and read the results from the graph. Check by solving the equations.

There are several values in such a procedure. First, approximate solutions can be obtained. Second, the effect of the constant terms is shown by noting the position of beginning the two lines when $d = 0$. Third, the effect of variation in the coefficient is observed by noting the slope or steepness of each line with respect to the horizontal. Fourth, the idea of dependence is expressed concretely. Fifth, interest is increased perceptibly.

(*h*) *Work Problems.*

With slight variation, all work problems can be attacked through a rather definite type of procedure. Consider the example:

> A man and his son can do a piece of work together in 8 days. The father does ⅓ more than the son. How long will it take the boy to finish if they work together 5 days?

The type is evident. Equality usually exists between certain parts of work done in a given unit of time, or between the units of time required to do certain amounts or parts of work. The pupil may then write:

$$\frac{1}{B}\text{, part done in 1 day by boy,} + \frac{1}{F}\text{, part done in 1 day by father,} = \text{part done in 1 day by both.}$$

B = no. of days it takes boy.
F = no. of days it takes father.

What is required? may be determined as t days for boy to finish, from which it may be written:

$$t = \frac{\text{part to be done}}{\text{rate, or part done in 1 day}}.$$

These equations have to be cleared by, What have I to find it with? Hence:

t = no. of days for son to finish.

$\frac{5}{8}$ = part done by both in 5 days.

$\frac{5}{B}$ = part done by boy in 5 days.

$\frac{5}{F}$ = part done by father in 5 days.

$$\therefore \frac{5}{B} + \frac{5}{F} = \frac{5}{8}.$$

But $\frac{1}{B} = \frac{2}{3} \times \frac{1}{F}$ since the son does $\frac{1}{3}$ less in 1 day. This last equation is necessary for eliminating one of the unknowns in the equation above.

$$\therefore \frac{5}{B} + \frac{3}{2} \times \frac{5}{B} = \frac{5}{8},$$

$$\text{or} \quad \frac{1}{B} + \frac{3}{2} \times \frac{1}{B} = \frac{1}{8},$$

$$\text{and} \quad t = \frac{\frac{3}{8}, \text{ part left}}{\frac{1}{8}, \text{ part done in one day}}.$$

(*i*) *Puzzle Problems.*

Puzzle problems are so varied and include so many types of operations that it is difficult to give suggestions on the method of their solution. If a method or scheme for solving each were given, it would perhaps defeat the purpose—that of arousing interest—for which they are selected.

In most of these problems it has been found helpful to suggest that the pupil translate the problem into algebraic short hand as it is read, and usually in the same order, beginning by putting down the unknowns as the problem dictated. To illustrate, take the old problem of the shepherd:

> A shepherd on being asked how many sheep he had, answered: "If I had as many more, half as many more, and seven sheep and a half, I should then have 500." How many had he?

Let the pupil begin by putting down in order what the shepherd had, x, then take dictation, putting x for "If I had as many more"; $\frac{x}{2}$ for "half as many more" and $7\frac{1}{2}$ for the remainder and then 500. The quantities will then be connected with the proper sign and the equation, $x + x + \frac{x}{2} + 7\frac{1}{2} = 500$ is formed.

Perhaps the digit problems prove most difficult. The pupil cannot grasp the significance of the coefficients, as in $100\,x + 10\,y + z$, in representing a number. In the beginning, assume $100\,x + 10\,y + z = 385$. Separate 385 in the same manner and write just beneath the general expression thus:

$$100x + 10y + z$$
$$100(3) + 10(8) + 5.$$

The comparison is simple and should remove the difficulty at once. Other difficulties encountered in solving problems of this type are peculiar to each of the several problems, therefore, a general discussion will not be given.

No effort has been made here to set up a strict classification of all problems as to types. It is recognized that in a more refined classification some of the divisions made may include more than one type. This procedure was chosen in the hope that one illustration might be made to cover more than one case. It is hoped, therefore, that the "transfer" may be made, and that some of the most urgent needs have been met.

After a type of problem has been established, and the general method of its solution developed, opportunity should be given the pupil to recognize a problem as to type when it is in a group of miscellaneous exercises. Frequent use of unclassified problems is urged, for it is through the analysis, choice of procedure, and graphical interpretation of relationships that ability is to be developed in the study of algebra.

Summary

From a study of teacher and pupil reactions to algebra problems, it is discovered that a large percentage of problems that experienced teachers considered important do not possess a great deal of interest for the pupil. Since interest is essential to sustained activity, it is deemed necessary that, in addition to giving a method of solution of problems, some means of maintaining interest should also be provided.

A survey of the literature on the teaching of algebra shows a dearth of what are considered real helps in the maintenance of interest in and the development of power through the solving of algebra problems. There are many helps of a general nature, in various books on the teaching of mathematics, but the teacher who is just entering the field needs more specific suggestions. Sources have been found in which various devices for solving problems have been given, but devices are dangerous in the hands of most inexperienced teachers. They are likely to be used mechanically by the teacher rather than made a means of showing relationships, and as an aid in the development of mental power.

The *analytic method* is generally conceded by teachers of mathematics to be the best method for the discovery of original solutions, particularly in difficult problems. Some kind of systematic approach to an analysis of problems, therefore, should be given the pupil whereby he will form a habit of asking himself significant, leading questions that will serve as a basis for analyzing each problem. The steps in such a scheme of analysis should be as few and as simple as possible. The plan outlined under "Problem Analysis" is easily retained by the pupil. It reduces the analysis to four significant steps, viz.:

1. The classification of problems as to types.
2. The discovery of equality in general between elements in each type.
3. The noting of what is required in each individual case.
4. The examination of given data.

When the pupil develops the ability to employ some such procedure, he is able to solve difficult problems.

Suggestions have been made, and instances pointed out, as to how algebra can be made to lead up to other branches of mathematics. This is done by introducing the notion of the increment as a means of determining slopes, by pointing out the slope of the curve as an indication of rates of change, and by showing the effect of the coefficient and of the constant term in a linear equation upon the direction and position of the line. The introduction of these notions can be made most easily through the use of the graph.

Problem analysis, as described above, together with the use of the graph in picturing relationships, trends, and rates of change, not only develop the ability to solve more difficult problems but provide the pupils with opportunity for greater mental activity.

APPENDIX B

ALGEBRA PROBLEMS SELECTED WITH REFERENCE TO INTEREST

The following list of problems has been compiled on the basis of their *interest* to the pupil. Where possible, problems were selected with reference to both *interest* and *importance*. The interest factor was given preference, however, and importance in many instances has been sacrificed for interest. No problem has been included that has been given a value in interest of less than 5, measured on a scale of 0 to 10.

The types under which the first 64 problems have been classified were chosen arbitrarily. Other types might have been added, or some that are given might well be combined under one title.

The second group of 42 problems has been placed purposely under the heading of *Miscellaneous Problems*, the reason being given in Appendix A, page 56. This group is divided on the basis of one and two unknowns.

The problems under each heading below are arranged in the order of probable difficulty—difficulty being based on the number of pupils out of 600 missing each problem.[1]

Age

1. A boy is now x yr. old. In what year was he born? (6), (0)
2. How many years must pass before your father is just twice as old as you are? (6), (0)
3. A father said to his son: "After 3 yr. I shall be three times as old as you will be, and 7 yr. ago I was seven times as old as you then were." What were the ages of father and son? (7), (0)
4. A teacher being asked his age responded, "At present I am twice as old as you. Ten years ago, I was 17 yr. older." How old are teacher and pupil to-day? (7), (0)
5. William's age is $\frac{1}{4}$ his age 20 yr. hence. How old is he now? (7), (0)

[1] NOTE: The numbers in parentheses at the end of each problem denote the Interest value assigned by the pupils and the Importance value as estimated by the teachers, respectively, for each problem. The scale of values ranges from 0 to 10.

Formula

6. Using the formula $s = \frac{1}{2} gt,$
$g = 32.2$ ft.
Find how long it will take an object dropped from an airplane 5,000 ft. high to reach the ground. (8), (5)
7. Write the formula for finding the cost of taxi riding in New York City if the rate is 15¢ for the first quarter mile and 5¢ for each quarter mile thereafter. How much would it cost to ride 5 mi.? If a passenger pays the driver 75¢, how far did he ride? (6), (8)
8. A boy wishes to find the cost per month of operating an electric toaster for 1 hr. each day. He finds these formulas in a science book.

$$W = V.A.$$
$$\text{Kilowatt} = 1{,}000 \text{ watts}$$

On his toaster he finds a plate saying 110 *V.* at 5 amperes. What was the cost per month at 10¢ per kw.-hr.? (5), (5)
9. Write a formula for finding the interest on $500 @ 6% for 6 yr., 9 mo., and 15 da. (6), (8)
10. A certain man is accustomed to tip a waiter in any restaurant 10% of his plate charge. Give formula of the amount of his bill. Find the amount of his bill for himself and wife if the plate charge is $1.75 each. (6), (8)

Indirect Measurement

11. Find the height of a tree, if, at 200 ft. from the foot of the tree, the angle of elevation is 45°. (7), (7)
12. Jim was flying a kite. When he had let out 500 ft. of string, the kite was directly above a garage 300 ft. from where he was standing. How high was his kite? (6), (7)
13. When an airplane is directly over point *A* an observer at point *B*, 460 yd. away, observes the angle of elevation to be 55°. How high is the plane? (7), (5)
14. In order to find the width of a river, a group of students laid off a straight line 500 ft. long from *A* to *C* along one bank. One of the group stood at *C* and located a point *B* on the opposite bank on a line at right angles to *AC*. Another student, standing at *A* measured angle *CAB* and found it to be 60°. How wide was the river? (8), (7)
15. Two boys, one 5 ft. 6 in. and the other 5 ft. 3 in. tall, wanted to measure a young tree they had planted 5 yr. before. They chose a spot so that when the first boy lay on the ground with his feet toward the tree, and the second stood at his feet, the first could just see the top of the tree over the second boy's head. They used a tape and found it was just 30 ft. from the first boy's head to the tree. How tall was the tree? (8), (7)

Lever

16. A teeter is 16 ft. long. Two boys, John weighing 85 lb. and George 76 lb., are trying to determine how to place the teeter so that they will balance. Where will be the point of support? (7), (6)
17. Some boys wanted to make a diving board, by resting a 10-ft. plank on a rock *F*, and holding down the end with stones. If *F* was about 2 ft. away from the stones, and the heaviest boy weighed 135 lb., what would be the minimum weight of stones necessary? (6), (7)
18. A man and a boy were to carry 400 lb. on a pole 9 ft. long. How far from the boy must the load be placed so that he shall carry one-fourth of the weight? (5), (6)
19. How heavy a stone can a man lift by exerting a force of 100 lb. with a crowbar 6 ft. long if the fulcrum is 18 in. from the stone? (5), (7)
20. A boy, who weighs 130 lb. finds that by throwing his whole weight on the end of a $7\frac{1}{2}$-ft. crowbar he can lift a large stone from the ground. He wonders how much the stone weighs. If the fulcrum is 1 ft. from the stone (neglecting the weight of the crowbar), what is the weight of the stone? (6), (6)

Mensuration

21. Two water mains with diameters of 18 in. and 24 in., respectively, are to connect with a larger main having a cross-section area equal to the combined areas of the other two. Find the diameter of the large main. (6), (5)
22. A man having 240 ft. of wire desired to fence a garden plot. He decided to make the width 20 ft. less than the length. How long should he make it? (5), (7)
23. A railroad company had to lay tracks around a mountain. The track when laid proved to be laid in an exact semicircle 11 mi. long. How many miles would they have saved if they had tunneled through the mountain instead? (5), (3)
24. John had a stick 5 ft. long. He wished to make a kite with one cross piece twice as long as the other. In what lengths would he cut the stick? (5), (8)
25. Some boys have a plot of ground 40 rd. wide and 50 rd. long. They wish to double the area of the lot. They find that they can make it only 10 rd. wider because of the road, which runs on one side. How much longer must they make it? (6), (6)

Mixture

26. If 4 gal. of a 50% solution of alcohol and water is put in a radiator, how much of the solution will it be necessary to drain out and add alcohol to make an 80% solution? (5), (6)
27. A man buys 2 gal. of 70% alcohol for his automobile. How much water must he add to have a 25% solution? (5), (8)

PERCENTAGE

28. If I buy stock at 150 which pays $9 a share annually in dividends, what per cent am I making on my investment? (5), (7)
29. The Commercial Club decided to sell programs at the football game in order to buy a mimeograph which cost $157.50. The programs cost 6¢ each. How many programs at a profit of 60% on the selling price must they sell? (5), (6)
30. The Federal Government and the State Government each levy a tax of 20% and 10%, respectively, on all prize fights. The amount received by both from the Dempsey-Sharkey fight was $292,680. Find the gate receipts. (6), (5)
31. A baseball team has won 30 games and lost 24. It has 16 games yet to play. How many of them may it lose and yet win 60% of the total? (6), (5)
32. What per cent profit does a girl make selling candy at 50¢ a pound, made from the following recipe? 1 lb. sugar, 7¢; ½ lb. chocolate 30¢ per pound; 1 can milk 5¢; allowing ¼ lb. for the weight of the milk not evaporated in cooking? (7), (7)
33. A school savings bank pays 3% interest compounded monthly. A city bank pays 3% interest compounded semiannually. How much more will a boy have to his credit at the end of the year if he deposits his savings of $1 a month in the school bank instead of in the city bank? (7), (9)

PROPORTION

34. Two men use ice from the same ice house. One takes a cake every 2 da., and the other takes a cake every 3 da. What should each pay if the cost of filling was $125? (6), (5)
35. John worked 3 nights after school for Mr. Smith and his brother worked 1 night. How should they divide the $5 that Mr. Smith gave them? (7), (10)
36. John sold stamps during his spare time to earn spending money. He asked his friend Henry to go into partnership with him. They agreed that John should receive 10% more of the profits than Henry, because John had started the business. If their profits for the first month were $7.42, how much should each receive? (6), (7)
37. Two boys made $67.50 one summer by taking passengers on a launch. The boy who owned the boat received twice as large a share of the profits as the other boy. How much did each receive? (7), (8)
38. A man who is 69 in. tall has a shadow 77 in. long. How tall is a tree whose shadow is 30 ft. long? (5), (6)
39. A teacher has been given a $5 bill to distribute as 3 prizes to the 3 best students in her class. If she wants to give equal amounts to the second and third, and double amount to the first, how much will each receive? (7), (6)

40. We wish to reduce a photograph for publication. The size is 5 x 7 in. and the width is to be reduced to 2 in. Find the length. (5), (6)

Puzzle

41. A shepherd on being asked how many sheep he had in his flock, said: "If I had as many more, half as many more, and $7\frac{1}{2}$ sheep, I should then have 500." How many sheep had he? (8), (0)
42. What number must be added to the numerator of $\frac{8}{21}$ in order to make the resulting fraction ? (5), (0)
43. The difference between two fractions is $\frac{1}{8}$. One fraction is $\frac{2}{3}$ of the larger. Find the fractions. (5), (1)
44. The sum of three consecutive numbers is 27. What are they? (7), (6)
45. I can think of two numbers whose sum is 23 and whose difference is 5. What are the numbers? (7), (3)
46. I know a number such that the sum of seven times the number and five times the number is 84. What is the number? (6), (4)
47. Separate 56 into three parts such that the second is four times the first and the third half the second. (5), (3)
48. How can a bill of $2.45 be paid with dimes and quarters by using the same number of each? (5), (4)
49. If 122 marbles were divided among 3 boys, so that the first had twice as many as the second, and the second had 6 more than the third, how many had each? (6), (0)
50. The length of a room is 3 ft. more than the width. If the length were increased by 3 ft. and the width diminished by 2 ft., the area would be the same. Find the dimensions of the room. (5), (0)

Time—Rate—Distance

51. A plane starts from New York to meet a west bound ocean liner then 200 mi. out. If the ship is making 30 mi. per hour and the plane makes 120 mi. per hour, how far out will the ship be when the plane flies over her, if the plane is flying with a tail-wind of 20 mi. per hour? (5), (5)
52. If a friend of yours who is driving an automobile at the rate of 18 mi. per hour has a start of 6 mi., at what rate would you have to drive to overtake him in 20 mi.? (5), (5)
53. A boy knows that his canoe can travel 5 mi. an hour downstream and 2 mi. an hour upstream. How far can he go and return, in 2 hr.? (5), (6)
54. A boy starts to paddle his canoe directly across a river at a rate of 5 mi. per hour. The current carries him down the river at the rate of 4 mi. What was the actual speed of his canoe? (5), (3)
55. A boy has 3 free hours on a spring afternoon. His father consents to drive him to the woods, which are 5 mi. distant, in the car at the rate of 25 mi. per hour. But the boy must walk back. He

walks 5 mi. per hour. Find how much time he can spend in the woods before starting back. (7), (6)

56. The "distance traveled" part of a speedometer on a truck is broken. The driver notices that in hauling a load of furniture from one town to another his average speed is 20 mi. an hour, while on his return trip it is 30 mi. an hour. The round trip takes him 1 hr. and 15 min. How far apart are the two towns? (5), (8)
57. A man is walking around the edge of a 1-mi. race track, at the rate of about 600 ft. per minute. If a horse on the track passes him at intervals of 7 min., what is the average rate of the horse? (5), (5)

Saving

58. Elizabeth wishes to earn this summer enough money to give her an allowance of 50¢ per week for 39 school weeks. Her father tells her he will give her half the amount she earns. How much must she actually earn? (7), (9)
59. What part of my weekly allowance of $2.00 must I save each week during the year to have $20.00 at Christmas? (6), (9)
60. A boy is putting his money in the school savings bank with the intention of buying a scout suit costing $4.75, when he has sufficient funds. If he already has $1.25 to his credit and deposits $.50 weekly, in how many weeks will he have enough to purchase the suit? (8), (7)
61. For every dime that a boy put in his bank, his father gave him a quarter. How much had the father given him when there was $52.50 in the bank? (6), (8)

Miscellaneous Problems Involving One Unknown

62. The distance from New York to Philadelphia is approximately 90 mi. An autoist travels to the latter city from the former in 3 hr. and 20 min. The return journey requires 5 hr. on account of heavy traffic. What is his average rate of travel? (5), (7)
63. A boy's average grade in three subjects is 88; what must his grade be in the fourth subject in order that his average in all four subjects shall be 90? (8), (9)
64. A father promised his son a gift of $200 on the day he would be twice as old as his son. If at the time of the promise the son was 9 yr. old and the father 33, how old would the son be when he received his present? (7), (3)
65. Your score on a certain kind of test is the difference between the number of questions you have correct and the number which are wrong. Find your score if there are n questions in the test and w of them are wrong. (8), (7)
66. Find the height of the high school, which casts a shadow of 20 ft. when the angle of elevation of the sun is 35°. (6), (5)
67. A boy wants to measure the height above ground of a certain limb

of a tree. He places a 12-ft. ladder against the tree directly under the limb and finds that it reaches a point 5 ft. from base of tree in a horizontal plane. What is the height of the limb? (7), (6)

68. On July 1, John bought a Ford for $370.00. The agreement was that each month he should pay one-twelfth of the principal plus the interest. What was the amount of the fifth payment? (6), (9)
69. How much money invested at 6% does a man need if he wishes to retire on an independent salary of at least $1,500 yearly? (5), (9)
70. The father of a boy at our school gave his son 48 ft. of old wiring and told him he could use it to fence in a plot of ground for a garden. Ralph wanted to make the plot as large as possible so he tried to find whether a circular or a square plot would have the greater area. Which kind did he decide to make? (7), (6)
71. What length is the base of the triangle of greatest area that can be cut out of a board 30 in. x 30 in., if the height must be 18 in.? If the board were cut in half, how would it affect the area of the triangle, and what about the base? (5), (5)
72. A man has a large tract of land but only 160 ft. along the road. He has 500 ft. of fencing. What are the dimensions of a plot he can fence along the road? (5), (7)
73. A high school track team needs four 12-lb. iron shots. The pattern shop and foundry classes agree to make the shots if the mathematics class will find the diameter. Find it. (7), (5)
74. A girl wishes to make a skirt cut on the bias of material 40 in. wide. How much will she have to buy if the edge of the skirt pattern measures $1\frac{1}{4}$ yard? (6), (8)
75. A boy has a 48-in. strip of picture molding 1 in. wide with which he desires to frame a picture. His picture is such that it needs trimming and he would like to have the picture within the frame two-thirds as wide as long. Into what lengths will he need to cut his molding if he miters the corners? (8), (7)
76. Out of 25 true and false questions, Mary got 19 correct. The next week she answered correctly 23 out of 30. Which was the better score? (7), (8)
77. In the chemistry laboratory John analyzes 2.08 gr. of milk chocolate for the amount of fat in it. He got the following weights: the empty flask weighed 34.21 gr., the flask plus the fat weighed 34.81 gr. Find the per cent of fat in the chocolate. (6), (8)
78. Mary wishes to save $75 for her vacation by July 1. If she starts saving the first of March, what per cent of her salary must she save if she earns $22.50 per week? (7), (9)
79. If John should buy 20 marbles and Henry should lose 10, John would then have four times as many as Henry, and three times as many as James. How many has John? (5), (0)
80. The difference of the squares of two consecutive numbers is 43. Find the number. (5), (5)
81. If $\frac{1}{12}$ part of a number be added to 9, the result is the same as if $\frac{3}{4}$ of the number were subtracted from 29. Find the number. (5), (0)

82. If John should buy 20 marbles, and Henry should lose 10, John would then have four times as many as Henry. How many has each? (5), (0)
83. A father said to his son, "For every day on which your grade in algebra is 85% or over, I will give you a dime, but for every day that your grade is less than 85% you must give me a nickel." After 15 da. the son had 90¢. For how many days was his standing above 85%? (7), (8)
84. A boy starts for a touchdown with an open field from his opponents' 40-yard line and runs at the rate of 100 yd. in 24 sec. Two seconds later an opponent starts after him running at the rate of 100 yd. in 20 sec. Will the first boy make the touchdown? (7), (0)
85. A boy has two hours to row upstream to a ball game. The rate of the current against which he must row is 3 mi. an hour, and the distance to the park is 12 mi. How fast will he have to row? (6), (5)
86. The average rate of driving of one autoist exceeds that of another by 7 mi. per hour. The first goes 175 mi. while the slower goes only 120 mi. Find the rate of each. (7), (5)
87. One factor which becomes exceedingly important in auto races is that the air resistance increases as the *cube* of the speed. If the resistance is M when a car is going 35 mi. an hour, how great was the resistance when Campbell went 206 mi. an hour? (7), (5)
88. Robert deposited money in the school savings bank $\frac{1}{4}$ as many times as Will. If Robert deposits 20¢ each time and Will deposits 10¢ each time, how many times had each made deposits when together they have $12.00? How much of this total belongs to Will? (6), (9)
89. One summer two high school boys made $367.50 by taking passengers on a launch. The boy who owned the launch was to receive $1\frac{1}{2}$ times as much of the profits as the other boy. How did they divide the profits? (6), (7)
90. Two boys do a piece of work in 8 hr. getting $5.00 for the job; how much should each receive, if one works one hour more than the other? (5), (7)
91. Two carpenters agreed to remodel a certain house for $950.00. The material used cost $185.00 and the remainder was divided between them so that the second received $\frac{2}{3}$ as much as the first. How much did each receive? (5), (5)
92. Mary and her sister wish to save $1.50 for the Red Cross. Mary wants to give 10¢ to her sister's 5¢. How much money must Mary save and how much must her sister save? (5), (7)

Problems Involving Two Unknowns

93. A boy has the same number of sisters as brothers; while his sister has one and one-half times as many brothers as sisters. How many sons and daughters are there in the family? (7), (0)

94. In a certain number of two digits the tens' digit is four times the units' digit and the sum of the digits is 10. What is the number? (5), (0)
95. The library staff of a certain school wished to raise $175.00 with which to buy a set of books. A public spirited citizen donated the use of his movie theater for a Benefit performance. The theater seats 800. How many adult tickets at 25¢ and children's tickets at 15¢ will they have to sell in order to realize the desired amount? (5), (7)
96. A confectioner wants to mix candy of the 60¢ grade and $1.00 per lb. grade in such a way as to sell the mixture for 95¢ per lb. What proportion of each should he use? (5), (5)
97. Mary asked Jane how old she was. Mary answered by saying, "I am now 8 yr. older than my brother. In 3 yr. I shall be twice as old as my brother is then. Can you tell how old I am now?" (7), (0)
98. The basketball coach announced in assembly that the boy's first team had won twice as many games this year as last year and two more. He also said that a total of 23 games had been won in the two seasons. Show what equations a boy might make and solve to find how many games were won each season? (6), (5)
99. A man decided to divide $20,000 between his two sons in such a way that the shares, which the sons receive at the age of 21 yr. accumulated at 4% compound interest, shall be equal. If the sons are 13 yr. and 17 yr. old, respectively, how much shall he give to each? (5), (9)
100. Margaret uses three times as much milk as water in making cocoa. How much of each will she need to make 3 qt. of cocoa? (5), (6)
101. If the numerator of a certain fraction be doubled and the denominator be increased by 5, the resulting fraction equals $\frac{7}{8}$. If the numerator be decreased by 1 and the denominator be increased by 7, the resulting fraction has the value $\frac{1}{3}$. What is the given fraction? (5), (0)
102. I am thinking of two numbers whose sum is 729 and whose difference is 341. What are the numbers? (8), (0)

APPENDIX C

ALGEBRA PROBLEMS SELECTED WITH REFERENCE TO THEIR IMPORTANCE

The following list of problems has been compiled with reference to their *importance*. In this case *interest* has been sacrificed for *importance,* if both factors were not present. These problems, ranked by a group of 500 trained and experienced teachers of algebra, have been given values in importance of not less than 5, measured on a scale of 0 to 10.[1]

While some of the problems could be solved by arithmetic, they serve as valuable material for practice in generalizing, for showing the dependence of one variable on another, and for picturing graphically relationship between quantities.

Averages

103. A player on a ball team, having been at bat 45 times, made 16 hits. What is his batting average? (8)

104. Two of the best players on the school baseball team kept an account of the number of hits they made when at bat. Robert made 115 hits out of 194 times at the bat and Henry made 98 hits out of 175 times. Which had the better batting average? (6)

105. The Giants have won twelve games and lost four. The Pirates have won ten and lost six. If beginning with to-morrow's game the Giants should win and lose alternately each day, and the Pirates should "hit a winning streak," how long would it be before the Pirates would be tied with the Giants? (7)

106. A boy's grade in arithmetic is 89. What is his grade in algebra if his average in arithmetic and algebra is 92? (5)

107. It is the custom of our principal not to give out regents marks. The students find that to determine his final mark the principal takes the average of the marks for the four quarters and averages this school average with the regents mark in the ratio of 3 to 1. What mark does the student receive in regents, if his quarterly marks are, respectively, 70, 78, 83 and 81, and his final average is 82%? (5)

[1] Note: The number in parenthesis at the end of each problem indicates the Importance value assigned by the teachers, measured on a scale of 0 to 10.

Formulas

108. To find Fahrenheit degrees from Centigrade degrees we use the following formula:

$$F = \frac{9}{5}c + 32. \qquad (5)$$

Find the value of the similar points on the two thermometers that are numerically equal.

109. Find the value of A in the formula $A = ab$ when $A = 12.32$ ft. and $b = 6.3$ ft. (8)

110. It has been found that the most pleasing proportions of a rectangle are given by the formula

$$\frac{W}{L} = \frac{L}{L + W}.$$

What should be the length of a banner which is to be 30 in. wide? (5)

111. A formula for the area of a circle is:

$$A = \pi r^2$$

How is the A changed if r is doubled? If r is divided by 2?

112. From an elevation h ft. high, a distance d mi. can be seen. A formula connecting h and d is:

$$4d = 7h.$$

Find the distance seen from an airplane 1,500 ft. high. (5)

113. A boy is sent to mail a package weighing 12 lb. to a city 250 mi. away. How much will the mailing cost him if the parcel post rates are as follows:

1st and 2nd zones up to 150 mi.	$C = W+4$
3rd zone, 150 to 300 mi.	$C = 2W+4$
4th zone, 300 to 600 mi.	$C = 4W+3$
5th zone, 600 to 1,000 mi.	$C = 6W+2$
6th zone, 1,000 to 1,400 mi.	$C = 8W+1$
7th zone, 1,400 to 1,800 mi.	$C = 10W+1$
8th zone, over 1,800 mi.	$C = 12W$ (6)

114. Write a rule for finding the number of seconds in a given number of hours. Express this rule as a formula. (5)

115. How much interest must a man pay if he borrows \$6,500 for 3 mo. at 6%? To pay the same interest (\$) on the same principal when the time is doubled, how is the rate affected? Solve for the rate in terms of the other letters in the formula.

$$T = prt. \qquad (8)$$

116. Show by a line graph the Thrift record of a certain school which deposited the following amount:

1st week	$ 5.20	
2nd week	7.75	
3rd week	8.00	
4th week	5.50	
5th week	7.25	
6th week	9.00	
7th week	10.50	(8)

117. Represent by a graph the standings of a boy in his algebra class during the months of September and October.

Weeks	1st,	2nd,	3rd,	4th,	5th,	6th,	7th,	8th,
Standing	80%,	72%,	68%,	70%,	60%,	88%,	90%,	94%.

118. The enrollment number in four branches of athletics, baseball, basketball, football and soccer, were found to be in the ratio 4:3:2:2. Draw a circular pictogram showing the relative enrollment and the most popular sport. (5)
119. Make a table of the amount of money you have spent in successive weeks and draw a graph. (5)
120. $S = \frac{1}{2}$ ft. is a physics formula which expresses the law of falling bodies. S is distance in feet, t the time in seconds, and g is the force of gravity. Using $g = 32$, draw the graph of this formula. From the graph determine how far a body will fall in 5 sec. (5)
121. Find graphically the meeting point of two autos starting from New Hope, Pa. toward New York. The one starts at 6 P.M., travels at average rate of 25 mi. per hour, and is held up $\frac{1}{2}$ hr. for repairs. The other starts at 6:45 P.M., travels at an average rate of 35 mi. per hour, and needs no repairs. (7)

Indirect Measurement

122. There are two telephone poles 24 ft. and 39 ft. high. If they are 39 ft. apart on level ground, what length of wire will be required to connect the tops of the poles? (5)
123. From one bank of a river the angle of elevation of the top of a cliff on the other side is 33° 19′. If the cliff is 120.2 ft. high, how wide is the river? (6)
124. At noon a 10-ft. pole cast a shadow 11.8 ft. long. What was the angle of elevation of the sun? (7)
125. Jim was flying a kite. When he had let out 500 ft. of string the kite was directly above a garage. How high was his kite? (5)
126. The height of a certain flagstaff is unknown, but it is observed that a flag rope fastened to the top of the staff is 32 ft. longer than the staff and that its end just reaches the ground when carried to a point 15 ft. distant from the foot of the staff. What is the height of the staff? (5)

Lever

127. Where will a 12 ft. plank balance if the weight on one end is 70 lb., and 50 lb. on the other end, neglecting the weight of the plank. (5)

127*a*. Two weights balance on a lever when one is 8 in. and the other 7 in. from the fulcrum. If 4 lb. is added to each weight they balance when, respectively, 16 in. and 15 in. from the fulcrum. Find each weight. (5)

128. A bar 10 ft. long weighing 2 lb. to the foot has a weight of 6 lb. attached to one end. How far from one end must the point of support be in order that the bar may balance? (5)

129. A beam of 10 ft. span is to support a load of 5 tons. Where should the load be applied, if one support is to carry $\frac{2}{3}$ of the load? (7)

130. A weight of 380 lb. is carried between two men by means of a pole. One man is 5 ft. from the weight, the other is 4 ft. How many pounds does each man lift? (6)

131. A lever is balanced on a fulcrum with a weight of 40 lb. at one end and 50 lb. at the other. If 5 lb. is added to the 40-lb. weight, the 50-lb. weight will have to be placed 1 ft. farther from the fulcrum to balance the lever. Find lengths of the two arms or the lever at first. (6)

Maximum and Minimum

132. If a farmer has 75 rd. of fence, how large a field could he fence off, so as to have the biggest pasture possible? (5)

133. What is the length of an edge of the largest cubical box that can be made from a piece of cardboard 20 in. square? (5)

134. Henry wishes to fence in the greatest amount of ground for his pets. His father has given him 120 ft. of fence. Should he place the fence in the form of a square, rectangle, or circle? (5)

135. What should be the dimensions of a box containing 1 cu. ft. capacity for packing chemical salts, such that the cost of constructing it would be a minimum. The cost of the material is to be $1.00 a square yard. (5)

136. The size of parcels which the post office will accept for transmission is limited by the regulation that the length and girth of the parcel when added together must not exceed 6 ft. Find the dimensions of the largest box with square ends which it is possible to send through the post office. (5)

Mensuration

137. A man wished to hang a picture from a molding with two wires. He could find but two pieces, one 16 in. long and the other 28 in. How much should he cut from one piece to add to the other to make the picture hang straight, allowing 1 in. for twisting the segments together? (5)

138. A baseball diamond is in the form of a square. How far is it from home to the second base? (5)

139. Find where to cut a board, which is 53 in. long, into two parts such that their difference is 25 in. (5)

140. If a room is 4 ft. longer than it is wide, and it takes 32 ft. of molding to go around it, how much floor space is there? (5)
141. My home state is about 200 mi. × 400 mi. What scale should I use to draw the map upon paper 8 in. × 12 in., which will be a simple scale to work with and still give the largest possible map? (7)
142. A workman wishes to enlarge a drawing of a rectangle, so that the area will be twice the original area, keeping the ratio of the length to the width unchanged. The original is 8 in. × 10 in. Find the sides of the enlarged rectangle. (5)
143. Two boys raised an aerial pole on top of their house. The pole was 20 ft. above the peak. The only place to which they could run a guy wire was the other peak 40 ft. away. How long was the guy wire? (5)
144. A man has a rectangular field which is 40 rd. long and 20 rd. wide. He wishes to increase the area by 300 sq. rd. If he can increase the width by only 2 rd., how much must he increase the length? (6)
145. How many pansies should one buy to make a border about a circular bed that is 10 ft. in diameter? The plants are to be placed 4 in. apart. (5)
146. Find the number of gallons a cylindrical pump will deliver in an hour. The diameter of the cylinder is $2\frac{1}{2}$ in. and its length is 10 in. It makes 50 strokes per minute. (5)
147. How many trays 1 in. long, w in. wide, and d in. deep must I fill with water to freeze in my electric refrigerator to make a 2-qt. freezer of ice cream, if 8 lb. of ice are required for the process and 1 cu. ft. of ice weighs 57.2 lb.? (5)
148. A girl has a piece of expensive lace which she wishes to put on a circular centerpiece. How may she find a way to cut a circular piece of fine linen in order to use all her expensive lace, but require no more? (5)
149. If air is a mixture of 4 pt. nitrogen to 1 pt. of oxygen how many cubic feet of each are there in a room 40 ft. × 30 ft. and 12 ft. high? (6)
150. The boys of a certain high school decided to have a blacksmith make their basketball rings. The official rule book stated that the inside circumference of the rings should be 18 in. What radius was the blacksmith instructed to use in making the rings? (5)
151. A bicycle wheel is 28 in. in diameter. How far will it roll in making one revolution? (5)
152. Find the quantity of sawdust required to cover the track around a circus ring if the diameter of the ring is 100 ft. and the track is 20 ft. wide? (5)
153. Two brothers are to mow their grass tennis court 36 ft. × 78 ft. How wide a strip must the older boy cut around the court to leave just half for his brother to do? (7)
154. A man wanted to move the hot water boiler from the kitchen of his house to the cellar, where there was a narrow space 15 in. wide.

Before he moved it he wanted to be sure that it would fit, so he measured its circumference and found it was 42 in. Did the boiler fit into the space? (5)

155. A roof was needed on a circular cistern 18 ft. in diameter. The peak of the roof was to be 6 ft. above the level of the top of the cistern. How long must the roof boards be to project 8 in. over the sides of the cistern? What is the pitch of this roof? Is it enough to protect the cistern fairly efficiently? How would the lumber be cut most efficiently and economically? (8)

156. Suppose that we were to make a square, uncovered box from a piece of sheet metal. The metal is square. Square corners are cut out. The remaining sides are bent up and then the edges are soldered. The box is to be 2 in. deep and is to contain 128 cu. in. How large shall the sheet metal square be cut, from which the box is to be made? (5)

157. Show the number of square yards in a square rod by a scale drawing. Let 1 in. equal 1 yd. Lay the drawing off in units of square measure. How many facts can we state about this drawing? What different areas are shown? How many of each? What is their total area in square yards? (7)

Mixture

158. A nurse has a 25% solution of silver nitrate, and the doctor calls for a 10% solution. How many ounces of pure water will she have to add to 20 oz. of the 25% solution in order to get a 10% solution? (6)

159. How much liquid must be evaporated from 100 lb. of a 5% solution of water and salt to make it an 8% solution? (5)

160. How much water must be added to 40 gal. of milk testing 5% butter-fat to make it test 4% butter-fat. (7)

161. A pound of a certain mixture contains 2 prt. by weight of graham flour to 3 prt. of white flour. How much graham flour must be added to this mixture to get a mixture which contains 3 prt. of white flour to 8 prt. of graham flour? (6)

162. How many ounces of gold must be melted with 20 oz. of gold 14 carats fine (14 pure gold) to make an ingot 18 carats fine? (5)

163. How many pounds each of 30¢ candy and 25¢ candy can I buy for 80c in a mixture of 3 lb.? (7)

164. An alloy of copper and tin contains 4 lb. of copper and 11 lb. of tin. How much tin must be added if we wish to obtain an alloy which is one-fifth copper? (6)

165. In 40 oz. of alloy for watch cases there are 15 oz. of gold. How much copper must be added to the alloy so that a watch case made from the new alloy weighing 2 oz. will contain ½ oz. of gold? (5)

166. If bell metal is made of 25 parts of copper to 11 parts of tin in weight, find the weight of each metal in a bell weighing 1,054 lb. (5)

167. A ticket seller wishes to change $15.00 into quarters, dimes, and nickels. He needs twice as many dimes as quarters, and three times as many nickels as dimes. How many of each will he need? (7)

168. A given mixture consists of 4 pt. of acid and 9 pt. of water. How many pints of water must be added to make the mixture 20% acid? (6)

169. How many cubic centimeters of an 80% alcohol solution should be added to 5 cu. cm. of 20% alcohol solution to produce a mixture containing 30% alcohol? (7)

Percentage

170. Two partners agree to divide their profits by allowing the one who owns the building 30% more than the other. How should they divide a profit of $12,000? (7)

171. How many years will be required for $6,000 to amount to $8,500 at 6% simple interest? (8)

172. If a certain kind of material shrinks 18% in laundering, how many yards must be bought to have 5 yd. after shrinkage? (7)

173. In Wyoming alone there are 38,000,000 acres of reserved and unreserved domain, accounting for close to 62% of the state's area. What is the area of the state? (5)

174. In grinding wheat into flour, there is a loss of 18%. How much wheat is needed to make 1,722 lb. of flour? (8)

175. A man buys a house for $10,000. If the upkeep of the property costs $600 per year, for how much a month should he rent it in order to receive 6% on the investment? (8)

176. In butchering a hog, 30% of its live weight is usually counted as by-products, the remaining part sold over the block in the meat shop. How many hogs averaging 210 lb. each must a slaughter house buy in order to deliver at least 15,000 lb. of meat to a meat shop? (8)

177. A man has 10 shares of stock bought at 125 which pays 6%. He also has 10 shares of stock bought at 86 which pays 5%. Which is the better investment? What per cent does he actually receive on each investment? (9)

Progression

178. A boy while in high school begins to save money to go to college. How much must he save each year if he increases his savings by one-fourth each year and at the end of four years has saved $576.56? (7)

179. Mr. Jones asks you to work for him for a period of 5 yr., with an initial salary of $660, and with a promise of a yearly increase of $50, if all is satisfactory. What would be your salary at the end of 5 yr.? How much would you earn during the time? Draw a graph and estimate your salary at the end of 8 yr., if you continued working. Is this true? (5)

180. A boy had an offer of two different jobs. The one paid $40 a month with a 50¢ increase each month, while the other started him at $30 a month and an increase of $1.00 per month. After what length of time would both jobs be paying the same amount? How much? Show graphically. (7)

181. If you were receiving $960 a year with no prospects of an increase and were offered another position at $800 for the first year and an annual increase of $50 a year for the next 10 yr., would you make or lose by accepting the proposition for a 3-yr. period? 10-yr. period? How much? (8)

Proportion

182. In the drawing of a picture of a tower which is 120 ft. high and 30 ft. in diameter, the diameter is represented by a line 5 in. in length. By how many inches should the height be represented? (5)

183. The number of posts required for a fence is 1500 when they are placed 15 ft. apart. How many would be needed if they were placed 10 ft. apart? (7)

184. In the United States nearly 20,000 children are killed every year, and for every death there are 26 serious injuries. About how many children have been injured during the past 2 yr.? (6)

185. Among the ingredients required to make a certain desert, the recipe calls for 3 eggs and 2 tablespoonfuls of flour. If I desire to reduce the quantity, using only 2 eggs, how much flour shall I use? (8)

186. Two families together purchased 75 shrubs for $21.75. How much should one family pay who took 45 of them? (8)

187. Three men have $17,000 invested in a business. The first $5,000, the second $7,500, and the third $4,500. If the profit for one year is $6,488.90, find each man's share. (8)

188. If a car travels 360 mi. in 2½ da., how long will it take it to travel 1,000 mi. (8)

189. If a man's earnings are directly proportional to the time he works, and if he earns $66 in 12 da., how long does it take him to earn $110? (5)

190. An excursion rate of 1⅓ fare for the round trip is advertised as $11.23. What is the trip rate, if full fare is paid? (7)

Rate–Time–Distance

191. The sound from a stroke of lightning was heard 8 sec. after the flash was seen. How far away was the stroke if sound travels at the rate of 1,080 ft. a second? (5)

192. A man leaves Norfolk, Va., at 10 A.M. for Richmond, Va., a distance of 135 mi. At what average speed should he travel in order to reach Richmond by 2 P.M., allowing 30 min. for tire trouble? (8)

193. If for every 50 ft. it rises vertically, an airplane takes 5 sec., how high will it go in 1 min. (5)

194. Mr. Smith leaves home at 9 A.M. in his car driving at the rate of

24 mi. per hour. At 11 A.M. his son starts out in his roadster and overtakes his father at 5 P.M. At what average rate did the son drive his car? (7)

195. Two men start at the same time and travel in opposite directions; the ratio of their rates is 2:3 and in 5 hr. they are 100 mi. apart. Find the rate of each. (6)
196. Two aviators start out from either coast of the United States. Their distance apart is 3,000 mi. The first travels at the rate of 110 mi. per hour and the second at the rate of 115 mi. per hour. In how many hours will they meet? (5)

Work

197. *A* can build a wall in 6 da. and *B* can do it in 8 da. How long will it take *A* to finish the job after both have worked 3 da.? (5)
198. A typist can copy a certain manuscript in 4 hr. Her assistant can do it in 6 hr. What part of the work can they do together in 1 hr.? (6)
199. *A* and *B* can do a piece of work in 12 da. After *A* has worked alone for 5 da., *B* finishes the work in 26 da. In what time can each alone do the work? (6)

Miscellaneous Problems in One Unknown

200. A farmer has a field of 20 acres which he wishes to plant in corn and potatoes in such a way that he shall have four times as many acres in corn as in potatoes. How many acres will he plant of each? (5)
201. A man has an opportunity to invest a sum of money in two different ways, one yielding 4% and the other 5%. What part of $10,000 should be invested at 4%, and what part at 5% to obtain an annual income of $450? (6)
202. A dairy receives an order for 20 gal. of cream which shall contain 25% butter-fat. To cream containing 40% butter-fat the dairy man wishes to add milk containing 8% butter-fat. How many gallons of each should be used? (7)
203. A week before "Mother's Day" a man sends a check for $5 to a florist in his home town and asks the florist to deliver to the mother a bouquet with roses and half as many snap dragons. If roses are $4 a dozen and snap dragons are $2, how many of each should the florist send? (6)
204. A farmer can buy Pure Bred cows for $500 each and Grade cows for $75 each. How many cows can he buy if he wants as many pure breds as his money will allow, and then uses the rest for grades when he has $4,725? (5)
205. A boat going 18 mi. per hour travels downstream from *A* to *B* in 2 hr. It then returns to a point *C* 7 mi. below *A* in 3 hr. What is the distance from *A* to *B*? What is the speed of the current? (5)
206. A crew which rows 8 mi. downstream in 40 min. requires 1 hr. to

row back to its starting point. What is the rate of the crew in still water? Rate of current? (6)

207. A merchant received a bill containing the following items—

June 7.	3 typewriters and 2 desks	$210.00
June 19.	2 typewriters and 4 desks	260.00
		$470.00

He wanted to know how much one typewriter cost. Can you tell him what he wants to know? (5)

208. A man wishes to invest $12,000, some to be invested in 4% securities and some in 5%. What is the greatest possible amount he can invest at 4% (which is the safer) in order to obtain at least $500.00 per year? (7)

209. The length and width of a certain landscape painting have the ratio of the Golden Section (0.6). The picture covers 42 sq. ft. of wall surface. What are its approximate dimensions? (5)

210. The committee for a church supper needs 10 lb. of coffee, for which they are willing to spend $3.50. The buyer finds that she can get good coffee at 28¢ per pound and a better grade at 45¢ per pound. How much of each will she buy? (7)

211. A school planned to give an entertainment for which the admission price for adults was to be 25¢ and for children 15¢. They wanted to take in $170. If the school auditorium could seat 800 people, how many of each kind of ticket would have to be sold in order to make $170? (5)

212. The proceeds from the sale of tickets for a bazaar were $35.00. The adults paid 75¢ and the children 50¢ each. How many of each kind were sold? (5)

213. Mr. Avery can plow a certain portion of ground with horses in 8 da. His son plowed a similar portion with a tractor in 2 da. They are planning to plow a field as large as both of those portions together. If they both work on this field, how long will it take them to complete the job? (5)

214. A road goes uphill for 7 mi. and then downhill for 3½ mi. at the same grade. A cyclist rides the entire distance in 1 hr. 14 min. He then returns over the same route in 58 min. What are his rates in miles per hour both uphill and downhill? (5)

215. An army cook wishes to prepare a meal that will contain 4% fat. He is to combine beef and beans containing 6% and 3% fat, respectively. How many pounds of each must he use in preparing 50 lb. of food? (7)

216. I sent a 31-word telegram for 98¢ and a 45-word telegram for $1.40. There is a fixed rate for the first 10 words and a different rate for each additional word. What are these two rates? (6)

217. At two stations *A* and *B* on a line of railway the prices of coal are $3.50 per ton and $4 per ton, respectively. If the distance between *A* and *B* is 150 mi. and coal can be shipped for ½¢ per

ton per mile, find the place on the railway between A and B at which it will be indifferent to a customer whether he buys coal from A or B. (7)

218. A merchant buys one kind of tea for 36¢ a pound, another for 60¢ a pound. He sells a mixture of the two for 44¢ a pound and gains 10%. In what proportion does he mix? (5)

219. When weighed in water, tin loses 0.137 of its weight and copper 0.112 of its weight. If a 20-lb. mass of tin and copper loses 2.39 lb., find the weight of the tin and the copper in the mass. (5)

220. In boring holes in a metal plate, 3 circles touching each other are to be drawn, the distances between their centers being .865 in., .650 in., and .790 in., respectively. Find the radius of each of the 3 circles. (8)

221. It is known that gold loses approximately $\frac{1}{19}$ of its weight in water and silver $\frac{1}{10}$ of its weight. I have a piece of metal—part gold and part silver. It weighs 20 lb. dry and 16 lb. when weighed in water. How much is gold and how much silver? (5)

222. A with grade homeroom club of 26 pupils has $21 to spend for a homeroom library. They wish to buy 26 books so that each may have one to keep at the end of the year. The two series of books that they are interested in are published in 75¢ and 90¢ editions, respectively. How many shall they buy of each series so that they will spend the exact amount in their treasury? (8)

223. A chemist has an acid in two forms, one 90% pure and the other 60% pure. How much of each must he take to make a mixture of 10 oz. 72% pure? (6)

224. A tea merchant finds that he is losing money. He believes that this is due to the fact that he is selling a grade of mixed tea for 50¢ a pound into every 60 lb. of which mixture he has put 25 lb. of 70¢ tea and 35 lb. of 40¢ tea. Is this the cause of his loss? How many pounds of each grade should be used? (6)

225. A farmer can buy ready mixed feed composed of corn and oats in the ratio of 2 lb. of oats for every pound of corn for $1.30 per cwt. By buying the grain at $1.50 per cwt. of corn and 90¢ per cwt. of oats could he increase the amount of corn per hundred weight of feed and still keep the price of his mixed feed the same? (8)

APPENDIX D

SUPPLEMENTARY ALGEBRA PROBLEMS FOR HIGH SCHOOLS: A MISCELLANEOUS GROUP IN ONE AND TWO UNKNOWNS

This list of 200 algebra problems is compiled with reference to interest on the same basis as those in Chapter VII.

These problems are not divided into types purposely. It is reasonable to think that the greatest development from the study of algebra comes through problem analysis. One phase of analysis is the determining of the type to which a problem belongs.

It is hoped, therefore, that this group will serve the purpose for which it is intended, as described in Chapter V.

Miscellaneous Problems

1. If you notice in the paper to-day that the closing quotation of a certain stock is $96\frac{1}{2}$ and that the net change is -3, what was yesterday's quotation? (9)
2. A woman has a piece of linen 18 in. wide. How long a table runner can she make if she has exactly 4 yd. of braid to go around it? (8)
3. Mr. Jones wishes to raise 250 chickens. He estimates that $\frac{5}{8}$ of all the eggs he puts into the incubator will hatch. How many eggs will be put into the incubator? (9)
4. The minimum temperature on a certain day in New York City was $-4°$; the maximum was $34°$. Find the range of temperature there on that day. (9)
5. A man has a farm consisting of 105 acres. How can he divide it between his two sons so that the older will receive twice as many acres as his brother? (6)
6. How much rope will be needed to reach from the top of a tent pole 10 ft. high to a point on the ground 7 ft. from the base of the pole? (5)
7. Housed dairy cows require 4 sq. ft. of light each; a dairy farmer has 30 cows; how many 4×3 windows are needed in the new barn he is building? (8)

8. What is the equivalent Fahrenheit reading to a reading of $-40°$ C.? (8)
9. $F = \frac{9}{5}C + 32$.
   (1) Change formula to one which shows how to find Centigrade (C.) degrees from Fahrenheit (F.) degree.
   (2) Change 66° F. temperature to the Centigrade scale. (8)
10. A baseball diamond is in the form of a square 90 ft. on a side. How far is it from home to second base? (7)
11. If I represents the interest, P the principal, and R the rate; write the formula for R. Use your formula and find the principal that will yield $100 at 5% interest. (10)
12. The cost of sending a package by parcel post into the sixth zone is 11¢ for 1 lb. and 9¢ for each additional pound or fraction of a pound. Write this statement as a formula and find the cost for a package weighing 12 lb. (10)
13. If my car travels 250 mi. in 1½ da., how long will it take me to complete an 800-mi. journey? (10)
14. The volume of a gas at 35° C. and 790 mm. pressure is 450 cu. cm. What is its volume at 90° C. and 630 mm. pressure? (5)
15. If a factory can turn out 328 machines in 8 da., how long should it take to fill an order for 205 machines? (6)
16. Counting 25 lines to the page and 6 words to the line, how many pages will be required to write an essay of 500 words? (8)
17. While in Canada we bought 15 gal. of gas at 29¢ per gallon. The gallon was called a "big gallon" measuring 5 qt. to the gallon. What was the price of gas at the same rate in the United States? (8)
18. A vertical stake set near a schoolhouse cast a shadow exactly its own length. Find the height of the schoolhouse if its shadow is 48 ft. long. (6)
19. The perimeter of a football field is 320 yd. and the length is 10 yd. more than twice the width. What are the dimensions of the field? (5)
20. Measure the diameter of an automobile wheel and determine how many revolutions it will make in going a mile. (5)
21. A baseball team has played 140 games during the season. It has won 38 games more than it has lost. How many has it won? What is its standing? (7)
22. A baseball is 8½ in. in circumference; what is its diameter? Solve by formula. (8)
23. The relation between ohms, volts, and amperes is expressed by the formula $A = V$. How many amperes will flow in a circuit in which the voltage is 200 and the resistance is 5 ohms? (5)
24. The formula used in a printing establishment to determine cost is $C = .10n + r$, where c represents cost, n is equivalent to the number of copies of printed material, and r indicates overhead expenses. If r equals $6, graph the value for c when different values are given n. (5)

25. In a catalogue the price of a sewer pipe is given by the formula $c = 0.4d + 14$. Where d is the diameter in inches, and c is the price in cents per foot of length, what is the cost of $20\frac{1}{2}$ ft. of pipe $3\frac{1}{2}$ in. in diameter? (8)
26. In 1926, the physical training director made a study of the heights of the boys in the first year high school. He asked one of the mathematics classes to get the results for him in as clear a form as possible. He gave them the following table. How did they compare these facts for him? (8)

Number	2	6	30	45	20	2	1
Height...........	45 in.	47 in.	50 in.	54 in.	56 in.	58 in.	60 in.

27. The annual budget of a family with an income of $1,500 was $350 for rent; $550 for food; $300 for clothes; $100 for recreation and vacation; $50 for fuel and lights; $50 for insurance and $100 for incidentals. Draw a circular graph representing these amounts. (7)
28. Show by means of a graph the relation between the number of births and deaths in your community during the past 10 yr. (9)
29. *BC* is the height of a lighthouse which is known to be 250 ft. From a ship at *A*, the angle of elevation is observed to be 12°. Find *AC*, the distance of the ship from the lighthouse. (6)
30. How long will it take an investment of $5,730 to amount to $6,589.50 at 3% simple interest? (10)
31. Ray drives a taxi. He charges 15¢ for the first $\frac{1}{2}$ mi. and 5¢ for each $\frac{1}{2}$ mi. after the first. Find by graphing the cost of a ride of 8 mi.; of 2.5 mi. (8)
32. A chicken fancier has a piece of land out of which he wants a rectangular lot 30 rd. long. He has an 80-rd. roll of wire and an extra part roll 26 rd. long which he wishes to use. How wide can he make the lot? (6)
33. I bought a trunk the dimensions of which are 32 in. × 17 in. × 21 in. I am now buying an umbrella which I wish to fit in on the bottom of the trunk. What is the length of the largest umbrella that I can use? (8)
34. John and Jack take care of Mr. Jones' garden for $80 for the summer, in order to earn money with which to buy a radio. They agree that John should have $1\frac{1}{3}$ times as much of the money as Jack, as he is to put more time on the job than Jack. How much will each receive? (5)
35. If the tax rate in a certain town is $2.25 on the $100, how much taxes would Mr. Allen have to pay if his property is valued at $15,000? (7)
36. A train goes from Pittsburgh to Philadelphia, a distance of 360 mi. in 8 hr. How long after the train starts will it pass through Horseshoe Curve which is 108 mi. from Pittsburgh? (7)
37. A man agrees to meet his brother who lives in a town 180 mi. away

at 6 P.M. on a certain day. He averages 30 mi. per hour traveling by automobile. At what hour must he start to be there on time? (8)

38. If A and B together invested $25 and $125, respectively, in a bond a year ago, and sold it yesterday for $225, how much should each receive? (8)
39. The cost of a macadam road was $36,000. The county paid twice as much as the state and the township three times as much as the state. How much did each pay? (8)
40. On a certain day in the winter in Boston, the lowest temperature reached was 7° below zero. On that same day in Miami, Florida, the lowest temperature reached was 70° above zero. What was the difference in temperature between the two places? (8)
41. Wednesday the thermometer registered —20° in the morning and 12° at noon. Friday it registered —17° in the morning and 13° at noon. Which day had the greater range of temperature and by how much was it greater? (8)
42. Two boys agreed to shovel off a certain amount of snow for $2.50. They found after finishing that one boy had shoveled a distance of 30 steps and the other a distance of 50 steps. How much money should each receive? (7)
43. Two families together purchase 75 shrubs for $21.75. One family took 45. How much should that family pay? (6)
44. A woman buys a piece of ice for 25¢. She finds it is approximately a rectangular solid which measures 12 in. × 10 in. × 6 in. Would it be more profitable to buy 50 lb. at 80¢ per 100? (5)
45. Mr. Martin lives so far from the center of the town that he cannot use the town water supply. He has decided to build a water tower on a low hill on his land. It is to hold 1,000 gal. of water when filled. If the diameter of its base is 12 ft., how tall must it be? (5)
46. A lawyer can afford to spend only $40 per week for office help. He must pay the office boy $5. He estimates the stenographer is worth two-thirds as much as the bookkeeper. How much should he pay each? (8)
47. What is the cost of maintaining ten 25-watt lamps burning 20 hr. at 10¢ per kw. hr.? (5)
48. A newsboy needs $9 to complete his Christmas purchases. If he makes $1.80 per week with 24 subscribers, how many subscribers must he have to make $3.00 per week? (8)
49. How many feet of molding must I purchase to go around a square room whose ceiling contains 576 sq. ft.? (9)
50. A boy scout, who can swim only a certain number of yards, comes to a river that he wishes to cross, but the distance, he thinks, may be too great. If you were the boy, how would you find the distance across? (5)
51. If your father placed $1,000 in your name on your first birthday, in a trust company that gave 5% interest and that allowed the interest to be added to the capital every year, find by using the

formula $A = P\ (1 + r)^n$ how much money you would need to continue your education at 18 yr. of age. (10)

52. In building a road the county pays twice as much as the state and the town one-half more than the state. What does each pay if the road costs $18,000? (8)
53. In making out his budget, Mr. Smith decided that he should only spend one-fifth as much for luxuries as he did for necessities, and that he should save twice as much as he spent for luxuries. How did he divide his salary of $3,600 among his luxuries, necessities, and savings? (8)
54. The speedometer, at the end of a day's auto trip in France registered 585 km. If the ratio of miles to kilometers is approximately 6 to 10, how long was the trip in miles? (5)
55. A girl spent a certain amount for pins, six times that sum for cloth, and one-half of the remainder for thread, needles, etc., and had 35¢ left. What was spent for each article? (8)
56. A lady wishes to make a circular flower bed 12 ft. in diameter. How many geraniums will she need to fill it, if she plans to place them 2 ft. apart? (8)
57. A boy in manual training making a bookshelf finds that the distance from the top of the lowest shelf to the underside of the top shelf to be 4 ft. 6 in. He desires to put between these shelves 4 other shelves of 1-in. board in such a way that the book space will diminish 1 in. for each shelf from the bottom to the top. What will be the several spaces between the shelves? (7)
58. An automobilist sets out on an 85-mi. trip. Since he must reach his destination within 5 hr., he drives at an average of 20 mi. an hour, in order to give himself a margin. After driving for 2 hr. he has trouble which delays him 1½ hr. What average must he maintain for the remainder of the distance in order to arrive on time? (8)
59. How far from a camp can a crowd of boy scouts hike if they go at an average rate of 5 mi. an hour, rest 2 hr. at lunch time, and return the same way at a rate of 3 mi. an hour in order to be back in 10 hr.? (7)
60. A certain class treasurer always planned to leave in the treasury at least $25 more than there was when he took over the position. Will you write a formula to express this? Draw a graph to represent this formula. (5)
61. The sum of three angles of any triangle is equal to 180.° The second angle is equal to three times the first and the third angle is equal to five times the first. Find the first angle. (5)
62. Find the altitude of a triangle with area 90 sq. ft. and base 20 ft. (5), (8)*
63. I wish to make a long distance telephone call where the charge is 75¢ for the first 5 min. and 20¢ for each additional minute. I am

* The second number after each problem indicates the importance of each as estimated by the teachers. The scale of values ranges from 0 to 10.

willing to spend $2.75 for the call. How long may I talk for that amount? (5), (8)

64. The City Hall of Philadelphia is said to cover a greater area than any other building in the United States. One-fifth of its width exceeds one-sixth of its length by 13 ft.; and one-ninth of its length exceeds one-tenth of its width by 7 ft. Find its dimensions. (7), (3)
65. A caterer prepares a dinner for 50 people at a cost of $50 to himself. How much must he charge per plate to be sure of making $17.80 if only 45 people come, counting the unused food as worth 20¢ per plate? (5), (7)
66. In certain states the highway commission will not allow public roads to be built with a gradient steeper than 1 ft. rise for every 10 ft. of roadway. What angle does such a road make with the horizontal? (5), (5)
67. A smoke stack is 65 ft. high. How long should an aerial wire be to reach from the top of the chimney to a window in a house 50 ft. away from the chimney? The window is 30 ft. above the ground. (5), (7)
68. A housewife wishes to make muffins for 3 people, while her recipe will serve 8. If it calls for 3 cups of flour and $\frac{2}{3}$ as much milk, how much of each shall she use for 3 people? (5), (8)
69. If the cost of gasoline varies directly with the number of gallons used and the number of gallons used varies directly with the distance, what will be the cost of gasoline for a 100-mi. trip if the cost for a 70-mi. trip is $1.20? (6), (7)
70. A father and a son contract to do a piece of work and agree to divide their earnings in the ratio of 5 to 3. They earn $40. How much does each receive? (5), (6)
71. The Traver Engineering Co. of Beaver Falls, Pa., are constructing an all steel coaster at Palisades Park. The weekly pay-roll for 95 men is $6,000. How many men will they have to discharge in order to keep the pay-roll down to $5,000 per week? (5), (6)
72. Colonel Lindbergh flew across the Atlantic Ocean in approximately 33.5 hr. at the rate of 108 mi. per hour. By equations find out how far approximately it is from New York to Paris by the route he took. (7), (5)
73. It takes a 250 ft. train 2 min. to completely pass over a bridge a mile long. What is the average rate of the train? (5), (5)
74. A farm is worked on shares. As the tenant supplies the tools and fertilizers, he receives twice as large a share of the profits as the owner of the farm. If the profits for one year are $6,000, how much does the tenant receive? The owner? (5), (8)
75. A man purchases an automobile for $800. He pays $200 cash and the rest in monthly installments of $40 and interest on the debt. If the rate of interest is 6%, find the amount of each payment and the amount of interest paid. (5), (9)
76. How much money at 6% does a man need to invest if he wishes to retire on an independent salary? (5), (9)

77. How many years would it take $1,000 to double itself at 6%? (5), (7)
78. How much does a stone weigh if a boy weighing 120 lb. can just balance it on a 10 ft. pole when the fulcrum is placed 18 in. from the stone? (7), (6)
79. A father earned twice as much as his son. The father worked 3 da. and the son 5 da., for which they received $39. How should the money be divided? (5), (8)
80. An automobile covers 5½ mi. in 15 min. At the same rate how long will it take to go 25 mi.? (7), (5)
81. On a certain map a distance of 6 mi. is represented by a line 1½ in. long. What distance will be represented by a line 7½ in. long? (5), (8)
82. A fleet of 16 trucks moves a quantity of dirt each day, each truck making one trip in 40 min. Bv speeding up the trips to take only 35 min. each, how many trucks may be released for other work, without cutting down the daily amount of dirt removed? (6), (6)
83. A man died and left his estate of $16,500 to his wife and three sons. His wife was to receive ⅓ of the total estate and each son was to get $500 more than his next younger brother. How much did each receive? (5), (8)
84. A man can do a certain piece of work in 6 days and a boy can do the same work in 9 days. When the man has finished half of the work, his employer finds that the task must be completed in another day. How many boys must he put to help the man in order to accomplish this? (5), (7)
85. Find the area of an isosceles right triangle whose hypotenuse is 32 in. (5), (8)
86. The New York law specifies that an apartment building may occupy only 80% of its lot. A contractor wishes to build a house 40 ft. by 160 ft. Find the area of the lot required by law. (5), (10)
87. The volume of a cylinder equals $\pi r^2 h$. How tall will the tinsmith have to make a quart measure that is 3½ in. in diameter? (5), (8)
88. A school girl's daily diet should contain 2,200 calories of which 12.5% should be protein calories. Approximately how many glasses of milk should she drink in order to obtain ¼ of her protein calories from milk. (note) 1 cup of milk contains 34 protein calories. (5), (8)
89. The rate of a river is 2.5 mi. per hour. Find the time required by a boat to go 25 mi. downstream and back, when its rate in still water is 4 mi. an hour. (7), (5)
90. If you have 8 oz. of a 10% solution of listerine and water, and wish to reduce it to a 4% solution, how much water must be added? (5), (8)
91. The truck travels on the average, 20 mi. an hour. We, in the car, travel 30 mi. an hour. Will we, in a 48-mi. run, be able to overtake the truck which left three-quarters of an hour before us? If so, in how many hours? (6), (6)

92. When Mr. Brown made his will he provided that his wife and daughter were to receive equal amounts and his son was to receive $5,000 more than either the wife or daughter. How was his estate of $65,693.30 divided? (5), (8)
93. Two boys wish to finish picking a certain cotton patch Saturday morning so that they may have the afternoon for themselves. If there are 19 rows left, and one boy can pick 3 while the other picks 2, how many will each have to pick before they are through? (5), (7)
94. By getting a 5% discount for cash, a merchant increases his profits 8% without increasing the selling price. What is his gain when he receives no discount? (5), (8)
95. Find the compound interest of $375.00 for 8 yr. at 4%, interest compounded semiannually. (5), (10)
96. Mr. Henry borrowed $10,000 from a bank to build a home. How much must he pay the bank every 6 mo. for the privilege of borrowing, if the bank charges 6% interest? (5), (8)
97. A man has $2,500 invested at 5%. How much must he invest at 8% to make his total income 6% of the total amount invested? (5), (8)
98. Suppose a man wants to have an income of $2,500 per year from money invested. How much money will he have to invest at 6% in order to get this amount yearly? (6), (10)
99. If you invest $100 now at 6% simple interest, how old will you be when it doubles itself? (6), (7)
100. How many feet does a boy save in running across a square field instead of running along two sides if the field is 50 ft. long? (6), (7)
101. What is the volume of a tank 14 ft. in diameter and 24 ft. high when three-quarters full? (5), (9)
102. An architect who has made plans for a house with a base 30×42 ft. finds that he must reduce the size. By what equal amount must he reduce the two dimensions of the house in order to make the area of the new base 925 sq. ft.? (5), (8)
103. Find the number of cubic inches of metal in a pipe 30 ft. long if the inside diameter is $1\frac{3}{4}$ in. and the outside diameter $2\frac{1}{2}$ in. (5), (7)
104. A track $\frac{1}{4}$ mi. long is to be built with parallel straight sides and semicircular ends. It is desired to have each of the straight sides 120 yd. long. How far apart will the straight sides have to be? (6), (6)
105. Two boys wish to make a concrete roller for their tennis court. Together they had money to buy 4 cu. yd. By measurement they had found that the length of a roller was about three times the diameter of one end. If the volume of the roller is equal to the product of the base by the altitude, and the area of the base is π times the square of the radius, what length roller could they make? What is the radius of the base? (5), (6)
106. Find the height of a building by making use of propositions. (6), (8)

107. Rates for advertisement are 25¢ a line the first day and 15¢ a line for each day after the first. How long can my advertisement stay in the paper for $3.00? (5), (7)
108. Two brothers are asked to mow a lawn which is 60 × 80 ft. Each boy is to mow one-half of the lawn. If the older brother begins his work first, how wide a strip must he cut around the lawn when mowing the grass to have cut half of it? (5), (8)
109. Water consists of 1 pt. of hydrogen and 8 pt. of oxygen. How many grams of hydrogen are contained in 100 gr. of water? (5), (7)
110. A boy has 25¢ with which to buy marbles. He thinks he should have one shooter for every 10 glassies. Shooters cost 2 for 5¢ and glassies cost 1¢ each. How many of each does he buy? (7), (5)
111. A cylindrical tank 20 in. long, base with radius of 5 in., is fixed so that it lies on its side. If the height of the liquid in the tank is 4 in., how many gallons are in the tank? (Use 231 cu. in. to 1 gal.) (5), (9)
112. What should be the radius of two equal pulleys with centers 12 ft. apart, in order that the connecting belt shall be 28 ft. long? (5), (8)
113. One spring a farmer fenced off 2 acres of an alfalfa field to be used as a hog pasture. What were the dimensions of the field? The next spring he decided to add 1 acre to the pasture. He wishes to leave the length of the pasture the same, how wide would the added plot be? (5), (8)
114. In how many years would a hotel become a paying proposition if its cost of construction and equipment was $1,160,000.00, the annual cost of upkeep is $400,000 with an average of 800 guests paying $3.00 per day? (Consider 360 days in a year.) (5), (8)
115. A stone dropped from the top of a precipice reaches its base in 5 sec. How high is the precipice? (7), (6)
116. Find the number of gallons a cylindrical pump will deliver in an hour. The diameter of the cylinder is 2½ in. and its length is 10 in. (5), (8)
117. I wish to excavate for a building 80 ft. × 200 ft. The excavation is to be 20 ft. deep. One steam shovel and necessary men and truck can remove 2,000 cu. ft. per day. I find that the company to lay the foundation for the building must begin work in 40 da. How many more steam shovels with necessary men and trucks must I employ to finish the excavation on time? (5), (8)
118. The rate of the current of a certain river is known to be 3 mi. per hour. Some boys in a motor boat found that it took them as long to go 8 mi. downstream as it did to go 3 mi. upstream. What is the rate in still water? (7), (5)
119. A man takes twice as long to row a certain distance against the current as with it. If he rows 9 mi. an hour in still water how fast does the current flow? (6), (5)
120. A man rows downstream at the rate of 6 mi. an hour and returns at the rate of 3 mi. an hour. How far downstream can he go and return within 9 hr.? (6), (5)

121. A certain recipe for fruit punch calls for twice as many oranges as lemons. If oranges are 35¢ a dozen and lemons are 25¢ a dozen, how many dozen of each must be used in making a punch to cost $4.75? (5), (8)

122. During the summer, the rope on the flagpole broke and pulled through the pulley at the top. The janitor destroyed the rope. It was necessary to determine the length of rope needed. A broom handle 5 ft. long was set in the ground and its shadow found to be 4 ft. The shadow of the flagpole was 24 ft. What was the length of rope needed? What would be the best way to get the rope through the pulley at the top of the flagpole? (6), (7)

123. John desires to make a garden out of a piece of ground three times as long as it is wide, and with a perimeter of 100 ft. He wishes to plan it so as to have a 1-ft. border of flowers around the edge, and the remainder in tomato plants. If he allows one plant for each square foot, how many plants will he need? (6), (7)

124. In an alloy of silver and gold weighing 40 oz. there are 5 oz. of gold. How much silver must be added so that 5 oz. of the new alloy shall contain only $\frac{1}{5}$ oz. of gold? (5), (7)

125. One firm offers a piano for $400 subject to discounts of 20% and 20%; another offers the same piano for $400 less discounts of 25% and 15%. Which is the better offer? (5), (8)

126. What is the highest price that Mr. Scott can pay for gloves which he wishes to sell for $2.50 a pair, if he wishes to realize a profit of 25%, figured on his selling price? (5), (9)

127. Water consists of 1 pt. of hydrogen and 8 pt. oxygen. How many grams of hydrogen are contained in 100 gr. of water, in which 10 gr. of salt have been dissolved? (5), (7)

128. An automobile traveling 20 mi. per hour is followed by one which starts from the same point 2 hr. later. At what rate must the second automobile travel in order to overtake the first one in 4 hr.? (7), (7)

129. Three men, *A, B,* and *C,* go into business together. *B* contributes twice as much capital as *A,* and *C* contributes three times as much as *A.* If the profit for a certain year is $12,000, what was each man's share? (5), (9)

130. Electricity costs 11¢ per kilowatt hour, (1,000 watts per hour) in New York. What is the cost of the electric current for lighting in a home which has six 50-w. lamps burning on an average of 1.5 hr. for a 30-day month? (5), (5)

131. *A* can do a piece of work in 3 da. and charges $5.00 per day. *B* can do the work in 4 da. and charges $4.00 a day. Will it be cheaper to hire *A,* or to hire *A* and *B* and let them do the work together? (5), (7)

132. A retail store sold merchandise regularly at a profit of 30% on the cost. During a special sale the store offered all merchandise at 20% discount off the regular selling price. What was the store's profit on sales of $10,400 at the special price? (5), (8)

133. If it takes a man one hour to row a boat a mile with the current going 20 mi. an hour, how long will it take him to row a mile against the current? How long will it take him to row a mile in still water? (6), (5)
134. A suit has a list price of $45. What will the dealer have to add to the list price if he wishes to give his customer a discount of 20% and still get about $45 for the suit? (5), (8)
135. A radio may be purchased for $125 cash or for $25 down and $20 a month for six months. What rate of interest is paid on the deferred payments? (5), (9)
136. A boy walked to the top of a mountain at the rate of 2 mi. per hour and down at the rate of 4 mi. per hour. How far is it to the top of the mountain? (6), (6)
137. *A* and *B* working together can do a certain piece of work in 6 da. It takes *B* 5 da. more than *A* to do the work. Find the number of days it will take each to do it alone. (5), (7)
138. If the radius of a circle is 10 ft., how much will it have to be increased in order that the area of the circle may be doubled? (5), (8)
139. In a certain taxing district a man owns property that has an assessed valuation of $5,000. He pays $57.50 half tax. Write the formula and find the rate in this taxing district. (5), (9)
140. A group of boys have a plot of ground 40 rd. wide and 50 rd. long. They wish to double the area and keep the length and width in the same proportion. Find the dimensions of the new plot. (6), (8)
141. I want to buy a house for $12,000. I only have $2,400, but find I can borrow $9,600 from the Building and Loan Society. This money is to be paid back in equal monthly installments for 10 yr. with interest at 6%. Find the amount I must pay back each month. (5), (8)
142. A man invests $750 at 4%. What sum would he have to invest at 7% in order to make 6% on the whole amount? (5), (8)
143. On a debt of $4,000 a man was paying 6% interest. At the end of each year, he plans to pay $300 on the principal, and also the interest on the balance of the debt during the year.
    (a) Write down his payments to be made at the end of years one, two, and three.
    (b) How many years will it take him to pay off the debt?
    (c) How much will he have paid by the time he has freed himself of the debt? (5), (9)
144. The Peoples Saving Corporation loaned me $7,000 with interest at 6% under the conditions that I pay $70 each month until the entire loan is paid. The interest is deducted each month and the rest of the $70 is to apply on the principal. How long will it take to pay off the loan? (5), (9)
145. Mary began teaching at the age of 21 at $1,400 a year. The first two years she received annual increases of $50, after that, $75. How much was she receiving when she was 32 years old? (6), (7)

146. Mr. K. borrows $2,500 at 6%. He agrees to pay his interest each year plus $500 on the principal. How much interest will he have paid when he is finally out of debt? (5), (9)
147. An insurance company is paying an annuitant $1,200 a year. The annuitant wishes to change to equal monthly payments. Figuring the discount rate at 3% what would be the monthly payment? (5), (9)
148. A display window is 27 ft. long and 9 ft. deep. The proprietor wants to double the area, but keep the window the same shape. How many feet must he add to the length and to the depth? (5), (6)
149. A father and a son were talking. Mr. Bruce, "Roy, you are a senior now and have had three years of mathematics. Tell me this. I want to purchase a triangular-shaped piece of ground 50 rd. × 70 rd. × 100 rd. Find for me the area of that plot of ground so I can offer him my price per square rod." (5), (7)
150. What sum of money invested at 4% compound interest from a child's birth until he is 21 yr. old will yield $1,000? Formula $A = P\ (1 + r)^n$. (5), (10)
151. The area of a field is known to be 162 sq. yd. If its length is twice its width, how many fence posts will be needed if they are to be 6 ft. apart? (5), (5)
152. A man desires to save a certain amount of money each year so that his infant son shall have $5,000 for a college education. The man has 18 yr. in which to accumulate this money. His bank pays 4% compound interest compounded quarterly. Assuming that the interest rate and period remains unchanged, how much must be deposited on the first of each year? (5), (9)
153. John has two opportunities to help work his way during his ninth year in school. One man offers a beginning wage of $25 with a monthly increase of $1; another offers a beginning wage of $20 and a monthly increase of $1.75. Which is more worthwhile? Draw a graph showing this. (7), (9)
154. The Bethlehem Trust advertised a Christmas fund in which you paid 10¢ the first week, 20¢ the next, 30¢ the next, etc., for 50 weeks, and got back $127.50. What was the interest? (5), (8)
155. How much higher must a village construct a cylindrical tank which is now 10 ft. in diameter and 10 ft. high in order to store 10,000 gal. of water? (5), (5)
156. A woman preserved 25 lb. of plums. If she used 2 lb. of sugar to every pound of fruit, how much sugar did she use? (5), (6)
157. Partnership Agreement of *A,B,C,* stated that profits and losses were to be shared in proportion to investments. *A* invested $9,000; *B* invested $18,000; *C* invested $27,000. The profits for the investments ending June 30, 1926, were $12,360.30. What was each partner's share? (5), (8)
158. How much water must be added to 6 pt. of a 25% solution of ammonia to reduce it to a 10% solution? (5), (8)

159. Where a railway line goes round a curve the outer rail is always raised above the inner rail a height given by the formula

$$h = \frac{wv}{1 - 25_r}$$

h equals elevation of outer rail in inches.
w equals width of the gauge in feet.
v equals greatest speed of train in miles per hour.
r equals radius of curve.
Solve this formula for *v*. What does the *uhv* formula describe? (6), (7)

160. Mr. Brown knows that he can sell a piece of property for $2,300. What is the maximum amount which he can pay for the property and still make 15% in selling it? (5), (8)

161. A man buys 50 baskets of peaches costing 20¢ per basket. How much must he charge per basket so that if he sells only 45 baskets he may still make 20% of the cost? (5), (7)

162. A certain boy is given $100 on his first birthday. On each of the successive birthdays, including his twenty-first one, the sum is to be increased $100. How much will he have received by the time he is 21 yr. of age? (7), (8)

163. I joined a Christmas Club and deposited in a bank 10¢ the first week, and each succeeding week I increased the deposit by 10¢. How much would I have in the bank at the end of the year? (5), (8)

164. If you join a Christmas Savings Club requiring a payment of 5¢ the first week, 10¢ the second, 15¢ the third week, etc., for 50 weeks, how much will you pay on the last payment and how much will you have deposited during the 50 weeks? (5), (8)

165. If a man deposits in a savings bank $5 on January 2; $10 on February 2; $20 on the March 2, how much will he have deposited by the end of the year? (5), (8)

166. A boy begins work with a salary of $12.00 per week; if he gets a 50¢ raise each week, what is his weekly salary at the end of the year? (7), (8)

167. If $2,500 is invested at 6% and $35,000 at 5½% how much must be invested at 8% to make the total income 7% of the total amount invested? (5), (8)

168. By obtaining a cash discount of 2%, a merchant increases his rate of profit on cost by 4% without increasing his selling price. What was his original rate of profit? (5), (9)

169. A man took out a 20-yr. endowment policy for $1,000 at age 23, the annual premium being $47.06, and lived to realize the endowment. If the same company charges $15.20 per annum for a 20-yr. term policy at age 23, what rate did the man realize on his investment. (6), (9)

170. A department store which sold on an average of 3,600 yd. of cloth a day, reduced the prices 10% and found that the daily cash

receipts from the sale of cloth increased 20%. How many yards were sold per day at reduced price? (5), (7)

171. What difference is made to the time of swinging of a pendulum 4 ft. long if it is shortened 3 in.? Prove your formula. (5), (5)

172. Collect data on one of the subjects below (or a better one that you think of) and draw the graph:
   1. Taxes on your home for the last 10 yr.
   2. Tax rate in your city for the last 10 yr.
   3. Temperature readings taken every hour from 7 A.M. to 7 P.M.
   4. Money spent last year in the United States on chewing gum, cosmetics, education. (4)

173. The angle of elevation of an airplane at a certain point is 61°4′. Find the height of the airplane, if a point 1,500 ft. away is directly below the airplane. (5)

174. What per cent are you above or below the normal weight as given in the health chart? (8)

175. A tree grew in 6 yr. from a height of 32 in. to a height of 20 ft. 3 in. What constant annual growth-factor would account for this increase? (5)

176. Write down the rule for finding the number of houses on the "odd-numbered" side of a street when you know the number of the last house on that side. (8)

177. If the price of gasoline is 25¢ a gallon, show by a graph the cost of any number of gallons up to 20 gal. (9)

178. A man bought 500 baby chicks for $90.00. How many broilers (2 lb. each) did he have to sell at 30¢ per pound to pay for the purchase of the 500? (7)

179. The staff of the bank learn from the Parks Department that the grounds secured for the annual picnic are 40 yd. wide and 80 yd. wide. The manager wishes to know whether the grounds are large enough to have the 10-yd. dash. The junior clerk, who had studied mathematics gave him the required information. What was this information? (7)

180. Refreshments for a high school class picnic cost $11.50. If there are 46 members in the class, how much should the total cost be increased if three members bring one guest each? (8)

181. A baseball nine has played 27 games of which it has won 18. It still has 9 games to play. How many of these must it win in order to bring its average of games won up to .75? (5)

182. A boy scout, who can swim only so many yards, comes to a river across which he is not sure he can swim. He has learned to sight an angle. If you were the boy, how would you find the distance across? (8)

183. How much will the taxi man charge you for an 8-mi. ride, if his rates are 15¢ for the first $\frac{1}{5}$ mi., and 5¢ for each additional one? (7)

184. In a certain city the water-rent is computed by meter measurement. Each householder is charged $3.00 a quarter for 5,000 gal. with an additional charge of 8¢ for every 100 gal. over the 5,000 gal. used

each quarter. Give formula for each householder's bill of water-rent. Find what would be the amount of water-rent one householder would pay if he used 30,000 gal. in a year. (6)

185. The fee for renting a cap and gown is $2.00. The cost of a cap and gown is $10.00. Ten per cent depreciation must be figured for wear and upkeep. How long before the cap and gown pays for itself? (5)

186. Mr. Johnson started on a 450-mi. automobile trip. It took him 7 hr. to make the first 165 mi. At this rate how long will it take him to finish the trip? (7)

187. A woman has 1 bu. of dirt to fill a porch box which is 36 in. long, 10 in. wide, and 8 in. deep. If there are 2,150 cu. in. in a bushel, how far from the top will the dirt come? (5)

188. A plumber wanted a cylindrical pipe of 7-in. diameter. He wanted it to lap 1 in. when fastened together. How wide a piece of galvanized zinc was needed? (8)

189. A man wishes to inclose a rectangular lot, the area of which is 1 acre, with chicken wire. He measures the width of the lot and finds it to be 121 ft. How many feet of chicken wire must he buy to inclose the lot? (7)

190. If two automobiles start toward each other at the same time from places which are 175 mi. apart, one traveling 22 mi. per hour and the other 28 mi. per hour and the other 28 mi. per hour, how soon will they meet? (6)

191. A farmer wishes to enclose a rectangular field for a pasture, making it 15 rd. wide. He wants to make it as long as possible, using 172 rd. of wire fencing, which he has on hand. How long can he make it? (7)

192. A man has $20,000 and must live on the income. At what rate must he invest his money if he needs $1,200 annually? (9)

193. How deep would a cylindrical cistern have to be made which is 4 ft. across, so that it will hold 50 bbl.? (8)

194. Divide $90 between A and B so that A receives $10 more than B. (9)

195. The people in a certain town plan to erect a Y.M.C.A. building at a cost of $260,000. Mr. Jones, a wealthy citizen, agrees to give one-fifth as much as is given by the citizens of the town. Mr. Smith, also a wealthy man, agrees to give one-tenth as much as is given by the citizens. In order to get the total amount of $260,000, how much must be given by the citizens of the town? (6)

196. If it takes 294 sq. in. of paper to cover a cube, what is the length of an edge of the cube? (7)

197. If 200 cu. ft. of air space are allowed in a schoolroom for each pupil, obtain a formula for the number of pupils (P) that can be accommodated in a room whose dimensions are l, w, and h ft. (9)

198. The length of a tennis court is 42 ft. more than the width, and the perimeter is 228 ft. Find the dimensions. (5)

199. Mary wishes to cover a box, except base, for her bedroom, with

cretonne. The dimensions of the box are 3 ft. $\times$ $1\frac{1}{2}$ ft. $\times$ $1\frac{1}{2}$ ft. Estimate number of square feet needed by finding the total surface to be covered. (6)

200. A man wants to change $58 into quarters, half-dollars and dollars. If he needs three times as many dollars as half-dollars, and half as many quarters as half-dollars, how many does he require of each? (7)

201. A farmer at auction sales of three farms made the following purchases: at the first farm he bought 3 cows, 5 pigs, and 2 horses for $475.00; at the second, 2 cows, 3 pigs, and 4 horses for $715; at the third, 4 cows, 6 pigs, and 3 horses for $680.00. A few months later he wished to sell all his live stock, and he desired to know what average price he paid for each animal. How would you find out, if you were the farmer? (6)

202. A man desiring to know what he had paid per ton for his coal the previous winter, found the following record in his account book:
Nov. 1—3 tons hard coal, 2 tons soft coal, $42/00.
Jan. 12—2 tons hard coal, 3 tons soft coal, $38/00.
If the coal was the same price at both times what did he pay per ton for each kind of coal? (6)

203. *A* and *B* working together can do a piece of work in 6 da. It takes *B* 5 da. more than it takes *A* to do the work. Find the number of days it will take each to do the work alone. (7)

204. The admission to a certain football game was 35¢ for members of the school at which the game was held and 50¢ for others. 800 tickets were sold for $325. Find the number of tickets sold at each price. (7)

205. James is helping at the candy counter at a school entertainment. He is sent out with a $10 bill to be changed into nickels, dimes, and quarters. He receives four times as many dimes as quarters and seven times as many nickels as quarters. How many coins of each kind did he bring back? (5)

206. I can purchase 6 cans of peaches and 12 cans of tomatoes for $1.68, or 12 cans of peaches and 6 cans of tomatoes for $1.96. What is the price per can of each? (7)

207. Six baseball teams wish to arrange a schedule of games in which each team shall meet every other team three times. How many games must be scheduled? (5)

208. The records of two workmen show that *A* can join the parts of 500 sink mats in 24 hr. *B* can do the same number in 30 hr. If these are the only two workmen available, what is the smallest number of working days of 8 hr. each in which the employing company can guarantee to finish 2,000 sink mats? (7)

209. How much alcohol must a druggist add to a pint of 5% solution of tincture of iodine to reduce it to 2%? (8)

210. A farmer has a field of hay which is 60 rd. long and 40 rd. wide. He wishes to cut half of it by moving around the field. How wide a strip must he cut? (8)

211. A box 2 ft. high is to have a content of 48 cu. ft. What should be its dimensions if the length is to the width as 3: *a*? (6)
212. A contractor has a walk to build which shall be 100 ft. long, 10 ft. wide, and 8 in. thick. How many cubic feet of cement, sand, and gravel will be required if they are used in the proportion of 1, 2, and 3, respectively? (7)
213. A barn is 100 ft. square. The rafters make an angle of 20° with the horizontal. The roof projects 2 ft. over the outside walls. If the cost of roofing is $.95 per square foot, what is the cost of covering the barn? (7)
214. Bill is making a bookcase. He is now ready to put in 2 shelves each 1 in. thick. He wishes each space beginning at the bottom to be 3 in. wider than the one above it. Show where Bill has to put the 2 shelves if the inside measure is 36 in. (6)
215. A man earns $300 per month. He saves 20% of his salary, and at the end of each 6 mo. puts it on time deposit at 4%. How much will he have saved at the end of 5 yr. (interest included)? (8)
216. A farmer has 20 gal. of milk testing 5% butter-fat. How many gallons of water must be added in order that the mixture shall test 4% butter-fat? (7)
217. The length of a rectangular field is three times its width, and a fence surrounding the field is 248 yd. in length. Find the length and width of the field. (6)
218. A crew can row 8 mi. downstream and back again in $4\frac{4}{5}$ hr.; if the rate of the stream is 4 mi. an hour, find the rate of the crew in still water. (6)
219. The mixture of water and alcohol in a 3-gal. automobile radiator was found to test 30% alcohol. How much of the mixture must be drawn off, and how much water added to reduce the mixture to 20% alcohol? (7)
220. A man placed $1,000 in a bank at 6% interest compounded annually. At the end of 3 yr. how much has he in the bank? (10)
221. There were 2,500,000,000,000 tons of coal in the United States in the year 1910 and 3,000,000,000 tons were consumed between 1900 and 1910. If the consumption of coal should double every decade, calculate, to the nearest decade, how long the coal in the United States would last. (6)
222. One boy beginning work just before entering high school starts at a salary of $40 a month and is promised a raise of $2 a month each year. Another starting 4 yr. later, after graduation, begins at a salary of $45 a month with a raise of $5 a month each year. In how many years will their salaries be equal? (8)
223. A beam 12 ft. long supported at each end carries a load of 3 tons at a point 5 ft. from one end. Find the load in tons (excluding the weight of the beam itself) on each support. (6)
224. A farmer wished to make an outlay of $10.00 per acre for grass seed. He planned to use $1\frac{1}{2}$ bu. per acre. One kind of seed cost $6.00 a bushel and the other kind, of which the mixture was to

be made, cost \$8.00. How many bushels of each kind did he use? (8)

225. A dairyman had cream which tested 85% butter-fat and milk which tested 10% butter-fat. How much milk must be added to the cream to make 500 lb. of milk testing 50% butter-fat? (7)

226. A grocer has walnuts at 35¢ a pound and pecans at 50¢ per pound. He wishes to make a mixture of 100 lb. which he can sell at 40¢ a pound. How many pounds of each must he use? (7)

227. I wish to obtain an average rate of 6% on the investment of \$3,000. I wish to invest some money in business at 8% and, for safety, the remainder in government bonds at 4%. How much should I invest in each? (9)

228. Find the length and width of a rectangle whose area is 891 sq. in., and whose length is 6 in. more than the width. (7)

229. A farmer mixes some millet and peas using $1\frac{1}{2}$ times as much millet as peas. After he has finished drilling he discovers that he has 45 bu. of the mixture left. At what price will he sell the mixture to his neighbor if the millet costs \$2.00 per bushel, and the peas \$1.10 per bushel? (8)

230. A boy had a piece of tin 20 in. long and 8 in. wide. He can make an open box from it by cutting out equal squares from the corners and turning up the sides. How large should these squares be in order that the box may contain 129 cu. in.? (7)

231. It is important for mother to know how much ice to order for the week. 50 lb. will melt in 3 days if the thermometer is 80° F., and in 2 days if the thermometer is at 90° F. For five days the temperature is steady at 85° F. The other two days it is at 90° F. The box holds 50 lb. at a time. How much ice is ordered? (8)

232. A wholesale agency buys belts rolled into coils. To avoid unrolling the coil, the firm wishes a formula to calculate the length of the belt in the coil. (6)

233. In how many different ways can the sum of \$5.10 be paid with half-dollars, quarter-dollars, and dimes, so that the whole number of coins used shall be 20? (5)

234. A man wishes to find the weight of a uniform bar 12 in. long by balancing it on a knife edge. He found that when the bar was supported 8 in. from one end, the system was kept in equilibrium by hanging a 10-lb. weight 1 in. from the other end of the bar. What is the weight of the bar? (5)

235. If your name is one of a thousand who have taken chances on 10 articles, what are the odds against your winning one of those articles? (5)

236. An electrical engineer needed a piece of copper wire having a given electrical resistance. He calculated that if the wire was of a given length and the diameter was equal to one-half this length, the required resistance would be obtained. The same resistance would also be obtained if both length and diameter would be halved. What is the relative amount saved by using a wire having the latter dimensions? Cost of wire varies as the volume. (6)

237. A boy wishes to ascertain the height of a tower. He knows that it is 31 ft. 6 in. from his window to the pavement below, and that the distance through which a body falls varies as the square of the time of falling. He drops a marble from his window and finds that it strikes the pavement in 1.4 sec. Then throwing a stone upward he observes that it takes just 3 sec. for it to descend from the top of the tower to the ground. What is the height of the tower? (7)

238. A mixture in an automobile radiator, with a capacity of 4 gal., contained 2 qt. of alcohol mixed with the water. After a week's driving, measurement showed that 3 pt. of liquid had evaporated. There was a sudden drop in temperature and the mixture had to be made up to 20% of alcohol, with the radiator full. How much alcohol would be added to make just the right mixture assuming that there was 10% greater evaporation of alcohol than of water from the original mixture? (7)

239. A sled after reaching the bottom of a hill moves 27 ft. the first second and $1\frac{1}{2}$ ft. less each succeeding second. In how many seconds will the sled stop? How far from the bottom of the hill does it go before stopping? (5)

240. Find the vertical and horizontal components of a force of 600 lb. if the angle which it makes with the horizontal is 24°. (5)

241. A group of 21 persons wish to meet each week, five at a time for social tea at each others homes, so long as they could do so without forming exactly the same group on any two occasions. How many meetings will they have? What will be the most times that any one hostess will have to entertain? (5)

APPENDIX E

IMPORTANT FORMULAS

The use of the formula in elementary science, in business, and in mensuration makes it important that every individual should know something about it. The following list was included in the study, and each formula evaluated by the group of teachers that ranked the problems.

All of the formulas included in the list were given values of importance of 5 or above, on a scale of 0 to 10. They are included here as a supplementary list to be used by the teachers of algebra.

Subject		Formula	
1. Acceleration of a moving body[1]	a	$= \frac{v_2 - v_1}{t}$	(5)
2. Area of a rectangle	A	$= lw$	(8)
3. Area of a parallelogram	A	$= bh$	(8)
4. Area of a triangle	A	$= \frac{1}{2}bh$	(7)
5. Area of a trapezoid	A	$= \frac{1}{2}h(b + b')$	(7)
6. Area of a square	A	$= s^2$	(10)
7. Standard Time east or west of here[1] ..	T	$= t \pm n$	(6)
8. Batting average of a baseball player[1] .	A	$= \frac{H}{B - w}$	(5)
9. Work	W	$= Fd$	(6)
10. Interest by 60-day method	i	$= \frac{p}{100}$	(9)
11. The number of hours a child should sleep[1]	H	$= 8 + \frac{1}{2}(18 - n)$	(7)
12. Time for cooking roast of beef[1]	T	$= 15p + 20$	(7)
13. Number of single desks for schoolroom[1]	N	$= \frac{lw - 2(li + wi)}{A}$	(6)
14. Percentage grade of pupil[1]	p	$= \frac{2C + E}{3}$	(8)
15. Average rate of travel[1]	R	$= \frac{2rr'}{r + r'}$	(7)
16. Perimeter of a square	p	$= 4s$	(7)
17. Perimeter of a rectangle	p	$= 2(l + w)$	(7)
18. Perimeter of an equilateral triangle ...	p	$= 3s$	(7)

[1] **Note:** Definition of terms at the close of the chapter.

Subject		Formula	
19. Single discount in terms of two successive discounts, a and b	c	$= a + b - ab$	(8)
20. Board feet in a piece of lumber	$B.F.$	$= \frac{L \times W \times T}{12}$	(8)
21. Velocity of sound in terms of temperature	v	$= 1{,}090 + 1.14(t - 32)$	(7)
22. Circumference of a circle	C	$= 2\pi r$	(8)
23. Area of a circle	A	$= \pi r^2$	(8)
24. Area of a polygon	A	$= \frac{1}{2}ap$	(5)
25. Total area of a rectangular solid	T	$= 2lw + 2hw + 2lh$	(6)
26. Distance between a gun and the target [1]	d	$= \frac{t(v + v')}{2}$	(5)
27. The number of plants for flower-bed [1]..	N	$= \left(\frac{w}{s+1}\right)\left(\frac{1}{s+1}\right)$	(7)
28. Tons of hay in round stack [1]	T	$= \frac{4c^2h}{2{,}000}$	(7)
29. Total area of a cube	T	$= 6s^2$	(8)
30. Lateral area of a right circular cylinder.	S	$= 2\pi rh$	(8)
31. Total area of a right circular cylinder ..	T	$= T\pi r(h + r)$	(8)
32. Lateral area of a cone of revolution ...	S	$= \pi rl$	(7)
33. Total area of a cone of revolution	T	$= \pi rl + \pi r^2$	(7)
34. Taxi fare in New York City, per person	C	$= \frac{R + 4R'(d - \frac{1}{4})}{N}$	(6)
35. Horse power of a steam engine [1]	$H.P.$	$= \frac{P \times A \times V}{33{,}000}$	(5)
36. Area of a sphere	A	$= 4\pi r^2$	(8)
37. Volume of a sphere..................	V	$= \frac{4}{3}\pi r^3$	(8)
38. Volume of a right circular cone	V	$= \frac{1}{3}\pi r^2h$	(8)
39. Volume of a right circular cylinder	V	$= Bh = \pi r^2h$	(8)
40. Rate—time—distance................	d	$= rt$	(7)
41. Interest	i	$= prt$	(10)
42. Amount at simple interest	A	$= p(1 + rt)$	(10)
43. Amount at compound interest........	A	$= p(1 + r)^n$	(10)
44. Amount compounded k times per year.	A	$= p\left(1 + \frac{r}{k}\right)^{kn}$	(9)
45. The lever	w_1d_1	$= w_2d_2$	(7)
46. Thermometer	F	$= \frac{9}{5C + 32}$, or	(8)
	C	$= \frac{5}{9}(F - 32)$	
47. Height of blackboard above floor, by grades [1]	h	$= 25 \text{ in.} + \frac{3}{2}(g - 4)$	(6)
48. Elevation of outer rail of curved track [1].	E	$= \frac{4BV}{5R}$	(5)
49. Diagonal of a cube	D	$= e(3)^{\frac{1}{2}}$	(7)

Subject	Formula	
50. Energy of moving body	$E = \frac{1}{2}MV^2$	(5)
51. Velocity of stream at the bottom of a river, given the velocity (v) at the surface	$V = (v + l) - 2(v)^{\frac{1}{2}}$	(5)
52. Kinetic energy	$K = \frac{1}{2}mv^2$	(6)
53. The geometric mean	$m = (ab)^{\frac{1}{2}}$	(6)
54. Approximate square root[2]	$(N)^{\frac{1}{2}} = (s^2 + r)^{\frac{1}{2}} = s + \frac{r}{2s}$	(7)
55. Total area of a cylindrical tank open at one end	$S = \pi r^2 + 2\pi rh$	(7)
56. Quadratic formula	$x = \frac{-b \pm (b^2 - 4ac)^{\frac{1}{2}}}{2a}$	(8)
57. Falling bodies	$s = 16t^2$	(7)
58. Width of belt for driving engine of any horse power (H.P.)	$W_b = \frac{800\ H.P.}{v}$	(5)
59. Height of a vertically thrown ball[2] ...	$y = It - \frac{1}{2}gt^2$	(6)
60. Area of an equilateral triangle	$A = \frac{s^2(3)^{\frac{1}{2}}}{4}$	(8)
61. Area of the sector of a circle of n degrees	$A = \frac{n}{360 \times \pi r^2}$	(7)

[2] Explanation of terms in above formulas

1. a = acceleration.
   v = initial velocity.
   v = final velocity.
   t = time
7. T = time east or west.
   t = time here.
   n = number of time belt east or west of here.
8. H = number of hits by the player.
   B = times at bat.
   w = times walked.
11. H = number of hours a child should sleep.
   n = number of years old.
   $\frac{1}{2}$ = number of years added to 8 hr. sleep for each year the child is under 18 yr. of age.
12. T = time necessary to cook a roast of beef.
   p = number of pounds of beef.
13. N = the number of individual desks.
   l = the number of feet in length of the room.
   w = the width of the room in feet.
   i = the length in feet of the aisle that extends around the room.
14. P = percentage grade of the pupil.
   C = class work.
   E = examination mark.
15. R = average rate on entire trip.
   r = uniform rate of going.
   r = uniform rate of returning.
26. d = distance between gun and target.
   t = number of seconds between firing of gun and sound from target.
   v = rate or velocity of the bullet.
   v = velocity of sound at a given temperature.

27. N = the total number of plants.
w = width of bed.
l = length of bed.
s = distance between the plants.
28. T = number of tons of hay.
c = circumference of hay stack (at base, expressed in feet).
h = height of stack in yards.
35. $H.P.$ = Horse Power.
P = pressure in pounds per square inch.
A = area of piston in square inch.
V = velocity of piston in feet per second.
47. h = height of chalk ledge.
g = grade of room.
48. E = Elevation of outer rail in inches.
B = width of track in feet.
R = radius of the curve in feet.
V = maximum speed in miles per hour of the train.
54. N = the number.
s = the largest square root in N.
r = the remainder of N-s^2.
59. y = the height of the ball after t seconds.
I = the initial velocity of the ball.
t = the time in seconds.
g = the force of gravity.